GUIDELINES FOR
PROCESS SAFETY
FUNDAMENTALS
IN GENERAL PLANT
OPERATIONS

Publications Available from the
CENTER FOR CHEMICAL PROCESS SAFETY
of the
AMERICAN INSTITUTE OF CHEMICAL ENGINEERS

GUIDELINES FOR
PROCESS SAFETY FUNDAMENTALS IN GENERAL PLANT OPERATIONS

CENTER FOR CHEMICAL PROCESS SAFETY

of the

AMERICAN INSTITUTE OF CHEMICAL ENGINEERS

345 East 47th Street • New York, NY 10017

Copyright © 1995
American Institute of Chemical Engineers
345 East 47th Street
New York, New York 10017

Library of Congress Cataloging-in Publication Data
Guidelines for process safety fundamentals in general plant operations
     p.   cm.
  Includes bibliographic references and index.
  ISBN 0-8169-0564-9 : $120.00
   1. Chemical industry—Safety measures.   I. American Institute of
Chemical Engineers. Center for Chemical Process Safety.
TP149.G8437   1995
660'. 2884—dc20   94–14064
                                    CIP

5 4 3 2 1 1 2 3 4 5

CONTENTS

PREFACE

The American Institute of Chemical Engineers (AIChE) has a 30-year history of involvement with process safety and loss control for chemical and petrochemical plants. Through its ties with process designers, builders and operators, safety professionals and academia, the AIChE has enhanced communications and fostered improvement in the high safety standards of the industry. AIChE publications and symposia have become information resources for the chemical engineering profession on the causes of accidents and means of prevention.

The Center for Chemical Process Safety (CCPS), a directorate of AIChE, was established in 1985 to develop and disseminate technical information for use in the prevention of major chemical accidents. Over 80 corporations from all segments of the chemical and hydrocarbon process industries support the center. They select CCPS's projects, furnish the professional volunteers who serve on the various subcommittees responsible for implementing the projects and help fund the Center. Each subcommittee's activities are coordinated by a member of the CCPS staff.

Since its founding, CCPS has published many volumes in its "Guidelines" series, the proceedings of a number of technical meetings, and teaching materials to aid in integrating process safety into undergraduate chemical engineering curricula; cosponsored several international technical symposia; developed training courses; and undertaken research projects. Although most CCPS books are written for engineers in plant design and operations and address scientific techniques and engineering practices, several guidelines cover the subject of chemical process safety management. Successful process safety programs are the product of committed and active participation of managers at all levels, who apply a systematic approach to process safety as an integral part of operations management.

This Guidelines book represents an effort to present currently recommended fundamental safety practices. It is hoped that adherence to these practices will aid in the prevention of operating accidents in the chemical process industries or, at a minimum, in reducing the risks associated with plant operations.

ACKNOWLEDGMENTS

The American Institute of Chemical Engineers (AIChE) wishes to thank the Center for Chemical Process Safety (CCPS) and those involved in its operation, including its many sponsors whose funding and technical support made this project possible. Particular thanks are due to the members of the CCPS Process Safety Fundamentals Subcommittee who collaborated to write this book. The major portion of the volume was prepared by Raytheon Engineers and Constructors, Inc. The CCPS Process Safety Fundamentals Subcommittee managed the project and provided direction and assistance to Raytheon Engineers & Constructors.

The Raytheon Engineers & Constructors team was directed by Thomas O'Brien, with Edward Luckiewicz as the other chief author. Other authors included Alexander D'Antonio, Richard Berkof, Eugene Michaels (Electrical), Richard Getz (Piping), and Richard Hyneman (Instrumentation). Some order was imposed by the technical writer, Dorothy Trauberman. Particularly efficient contributions were made by Antoinette DeChristopher, who handled the word processing, and Marie Knup, company librarian. The CCPS Process Safety Fundamentals Subcommittee was chaired by Arthur Wildman of Merck and Company. Its current and most recent members are Urs Keller, Ciba-Geigy Corporation; Harry Weinberg, Rohm & Haas Delaware Valley, Inc.; and Wayne Zarnecki, Nalco Chemical Company. Former members are Paul Kremer, PCR, Inc.; Paul Krueger, BOC Group; Frank Manley, Rohm & Haas Delaware Valley, Inc.; Arthur Miller, Eastman Kodak Company; and Robert Pitrolo, Union Carbide Corporation. The Subcommittee wishes to acknowledge the assistance and suggestions of the following peer reviewers: Daniel Crowl, Michigan Technological University; Mark Eidson, Stone and Webster Engineering Corp; Peter Fletcher, Raytheon Engineers and Constructors; Thomas Gibson, Dow Chemical Company; Peter Lodal, Tennessee Eastman Corp.; David Moore, Acutech Consulting; and Charles West, PPG Industries, Inc.

The members of the Subcommittee and the authors wish to thank their employers for providing time and support to participate in this project.

Lastly, the Subcommittee also wishes to express its appreciation to Thomas Carmody, Bob Perry, and Felix Freiheiter of the CCPS staff who were responsible for the overall administration and coordination of this project.

ACRONYMS AND ABBREVIATIONS

ACGIH	American Conference of Governmental and Industrial Hygienists
AGMA	American Gear Manufacturers Association
AIChE	American Institute of Chemical Engineers
AISI	American Iron and Steel Institute
AIT	Autoignition temperature
AMCA	Air Movement and Control Association
ANSI	American National Standards Institute
API	American Petroleum Institute
ARCHIE	Automated Resource for Chemical Hazard Evaluation
ASHRAE	American Society of Heating, Refrigeration and Air Conditioning Engineers
ASME	American Society of Mechanical Engineers
ASTM	American Society for Testing and Materials
AWS	American Welding Society
AWWA	American Water Works Association
BLS	Bureau of Labor Statistics
BMCS	Bureau of Motor Carrier Safety
CAER	Community Awareness and Emergency Response
CAS	Chemical Abstracts Service
CDA	Copper Development Association
CEFIC	European Chemical Industry Council
CEMA	Conveyor Equipment Manufacturers Association
CERCLA	Comprehensive Environmental Response, Compensation and Liability Act
CFR	Code of Federal Regulations
CGA	Compressed Gas Association
CGI	Combustible gas indicator
CHEMTREC	Chemical Transportation Emergency Center
CHRIS	Chemical Hazards Response Information Service
CHLOREP	Chlorine Emergency Plan

CI	Chlorine Institute
CIP	Clean in place
CMA	Chemical Manufacturers Association
CPC	Chemical protective clothing
DCS	Distributed control system
DHHS	Department of Health and Human Services
DOA	Department of Agriculture
DOE	Department of Energy
DOL	Department of Labor
DOT	Department of Transportation
ED	Effective dose
EMI	Electromagnetic interference
EPA	Environmental Protection Agency
ESS	Emergency shutdown system
FDA	Food and Drug Administration
FEMA	Federal Emergency Management Agency
FHWA	Federal Highway Administration
FID	Flame ionization detector
FIFRA	Federal Insecticide, Fungicide and Rodenticide Act
FMEA	Failure Modes and Effects Analysis
FMECA	Failure Modes, Effects and Criticality Analysis
FMS	Factory Mutual System
FR	Federal Register
FRA	Federal Railroad Administration
FRMAP	Federal Radiological Monitoring Assistance Plan
FRP	Fiber-reinforced plastic
FTA	Fault Tree Analysis
GC	Gas chromatography
GPO	(U. S.) Government Printing Office
HACS	Hazard Assessment Computer System
HAZCOM	Hazard communication (per 29CFR1910.1200)
HazOp	Hazard and Operability Analysis
HSWA	Hazardous and Solid Waste Amendments
ID	Identification
IDLH	Immediately dangerous to life or health
IEEE	Institute of Electrical and Electronics Engineers
IR	Infrared
IS	Intrinsic safety
ISA	Instrument Society of America
JIT	Job-instruction training
JSA	Job safety analysis
LC	Lethal concentration
LD	Lethal dose
LEL	Lower explosive limit

LFL	Lower flammability limit
LPG	Liquefied petroleum gas
MAC	Maximum allowable concentration
MACT	Maximum achievable control technology
MESA	Mine Enforcement Safety Administration
MHI	Material Handling Institute
MSDS	Material Safety Data Sheet(s)
MSHA	Mine Safety and Health Act
MTB	Materials Transportation Bureau
NAAQS	National ambient air quality standards
NACA	National Agricultural Chemicals Association
NACE	National Association of Corrosion Engineers
NDE	Non-destructive examination
NDT	Non-destructive testing
NEC	National Electrical Code
NEMA	National Electrical Manufacturers Association
NESHAPS	National emissions standards for hazardous air pollutants
NF	National Formulary
NFPA	National Fire Protection Association
NIOSH	National Institute for Occupational Safety and Health
NPDES	National Pollutant Discharge Elimination System
NPSH	Net positive suction head
NRC	Nuclear Regulatory Commission
NSC	National Safety Council
NSPS	New source performance standards
NTSB	National Transportation Safety Board
OHMTADS	Oil and Hazardous Materials Technical Assistance Data System
OJT	On-the-job training
ORP	Oxidation-reduction (or redox) potential
OSHA	Occupational Safety and Health Administration
OSHAct	Occupational Safety and Health Act
OSHRC	Occupational Safety and Health Review Commission
PCB	Polychlorinated biphenyl
PEL	Permissible exposure limit
PFD	Process flow diagram
PHA	Preliminary Hazard Analysis
PrHA	Process Hazard Analysis
PID	Photoionization detector; proportional–integral–derivative (control)
P&ID	Piping and instrumentation diagram
PLC	Programmable logic controller
PM	Preventive maintenance
PPE	Personnel (or personal) protective equipment
ppm	Parts per million

PSTN	Pesticide Safety Team Network
RCRA	Resource Conservation and Recovery Act
REL	Recommended exposure limit
RFI	Radio-frequency interference
RH	Relative humidity
RTECS	Registry of Toxic Effects of Chemical Substances
SAE	Society of Automotive Engineers
SAMA	Scientific Apparatus Manufacturers Association
SAR	Supplied-air respirator; Safety Analysis Report
SARA	Superfund Amendment and Reauthorization Act
SBA	Small Business Administration
SCBA	Self-contained breathing apparatus
SOCMA	Synthetic Organic Chemicals Manufacturers Association
SWMU	Solid-waste management unit
TCLP	Toxic chemical leaching procedure
TD	Toxic dose
TDH	Total dynamic (or delivered) head
TEFC	Totally enclosed fan-cooled
TEMA	Tubular Exchangers Manufacturers Association
TENV	Totally enclosed non-vented
TLV	Threshold limiting value
TLV-C	TLV—ceiling
TLV-STEL	TLV—short-term exposure limit
TLV-TWA	TLV—time-weighted average
TSCA	Toxic Substances Control Act
UEL	Upper explosive limit
UFL	Upper flammability limit
UL	Underwriters Laboratories
USCG	United States Coast Guard
USP	United States Pharmacopoeia
UST	Underground storage tank
UV	Ultraviolet
VOC	Volatile organic compound

GLOSSARY

Aerosol: A dispersion of microscopic solid or liquid particles in air. Solid aerosols are classified as **dust, fume** or **smoke.** Liquid aerosols are considered either **mist** or **fog.**

Asphyxiation: Death from lack of oxygen.

Autoignition temperature: The temperature above which spontaneous ignition of all or part of a flammable substance or mixture can occur.

Burn-in: The earliest period of the failure-rate characteristic curve for engineered equipment. This relatively high rate of failure arises from faults in manufacture or assembly not detected during factory or precommissioning tests.

Cavitation: Generation and collapse of gas bubbles, producing vibration and sometimes severe mechanical strain. A common phenomenon in pumps, with bubbles forming in the low-pressure suction and collapsing in the pump. Cavitation reduces pump capacity and increases wear on components that are in direct contact with the collapsing bubbles, such as impellers and piston heads. See **NPSH.**

Clo unit: The unit of thermal insulation required to keep a sitting, resting individual comfortable in a ventilated room at 70°F and 50% RH. Each clo unit compensates for a drop of about 16°F in surrounding temperature and provides a resistance of about 0.88 ft^2-hr-F/Btu.

Curie Point: Temperature of a material below which there is a spontaneous magnetization in the absence of an externally applied magnetic field and above which the material is paramagnetic (i.e., extent of magnetization proportional to strength of the field).

Conservation vent: A venting device which controls the release of vapors from a storage tank while preventing vacuum formation and overpressurization of the tank.

Cryogenic liquid: A refrigerated liquid having a boiling point below –130°F at atmospheric pressure.

Deflagration: Burning which takes place at a flame speed below the velocity of sound in the unburned medium.

Detonation: A flame front traveling through a flammable gas or vapor at a velocity greater than or equal to the speed of sound. A stable detonation travels at the speed of sound; an overdriven detonation travels at a higher velocity.

Detonation arrester: An arrester designed to prevent the propagation of deflagrations and detonations.

Dewar: A double-walled container for cryogenic liquids, with evacuated annulus and heat-reflective surfaces facing the vacuum.

Dust: Airborne solid particles, typically between 0.1 and 50 microns in size.

Effective dose (ED): The dosage at which the effects of a toxic material are noticeable but minor and reversible. The specific values are often plotted as dose–response curves or given as ED_{10} or ED_{50}. The latter is the effective dose at which symptoms are noticeable in 50% of the reported subjects. The values of the ED are always less than the toxic dose (TD). See **lethal dose** and **toxic dose.**

Ergonomics: An applied science that coordinates the design of devices, systems, and physical working conditions within the capacities and the requirements of the worker.

Explosive limits (LEL and UEL): The composition of a vapor–air mixture, expressed as volume percent of the vapor, at which the mixture is just able to ignite and burn. For any given vapor in air, there are two limits: the lower explosive limit (LEL) and the upper explosive limit (UEL). Below the LEL, the vapor–air mixture is too lean to burn. Above the UEL, the mixture is too rich. The terms lower flammability limit (LFL) and upper flammability limit (UFL) are synonymous with LEL and UEL.

Extrinsically safe: A term applied when safety is built in by adding instrumentation, controls, alarms, interlocks, equipment redundancy, safety procedures, and the like during engineering, design, construction, or operation of a component, system, or facility. Contrasted with **inherently safe.**

Fault-tolerant: Not disabled by a fault in any one component (of control systems).

Flame arrester: A device used to prevent the propagation of a flame front initiated on its unprotected side through the device into a vessel or piping system. A flame arrester is not necessarily suited for detonation wave suppression (see **detonation arrester**).

Flammability limits (LFL and UFL): See **explosive limits.**

Flash point: The lowest temperature at which a liquid can evolve enough vapor to form a flammable mixture in air. There are two methods commonly used for determination of flash points, the open-cup and the closed-cup methods. The open-cup flash point is usually slightly higher than the closed-cup value. The closed-cup flash point for hydrocarbons can be approximated by

$$T_F = 0.6813 T_B - 118.86$$

where T_F and T_B are, respectively, the closed-cup flash point and the boiling point of the hydrocarbon, both in degrees Fahrenheit. NFPA 30 recognizes different standard methods for different materials. Liquids with viscosities less than 45 SUS and flash points below 200°F should be tested by ASTM D56. Others should be tested by ASTM D93. There are also alternative methods which may be used.

Fog: A visible **aerosol** of a liquid.

Fretting: A form of corrosion that occurs from vibration of bearings, gears, and hubs pressing against a rotating machine component.

Fugitive emissions: Releases of vapor which occur continuously during normal operation of equipment. These include leaks from seals, packing, gaskets, process drains, etc.

Fume: A dispersion in air of volatilized solids which are usually less than one micron in size.

Fusible plug: A nonreclosing pressure-relief device which functions by yielding or melting a plug of material at a predetermined temperature.

Galling: The welding together of high spots on contact surfaces of machinery resulting from high loads and high temperature. Also caused by high loads with inadequate lubrication.

Gas: A formless fluid that will completely occupy an enclosure at normal temperatures and pressures. See **liquid.**

Hazard: A characteristic of a system, plant, or process that represents the potential for an accident.

Hazard and Operability study (HazOp): A technique for identifying hazards and operating problems by using a series of guide words to identify possible deviations from intended process parameters.

Inertion: Reduction of the oxygen content of a space by displacement of air with an inert gas. The process of inertion may use a continuous flow of inert gas while venting the space or may rely on stepwise replacement by alternately raising the pressure with the inert gas and lowering it by venting or application of vacuum.

Inherently safe: A term applied to a component, system or facility in which potential dangers have been removed rather than designed for. Inherent safety is incorporated during development, design, or engineering. Also see **extrinsically safe.**

Intrinsically safe: A term applied to the installation of electrical devices in an explosion hazard area. An intrinsically safe system is one designed, installed, and maintained so that under the most adverse conditions it cannot produce an arc of sufficient energy to be considered a source of ignition.

Lading: The contents of a container prepared for shipping.

Lethal dose: The dosage at which the effects of a toxic substance are deadly and irreversible. See remarks on dose–response characteristics under **effective dose** and **toxic dose.**

Liquid: A formless fluid that can partially occupy a container. See **gas**.

Lower explosive limit: See **explosive limits**.

Lower flammability limit: See **explosive limits**.

Maximum allowable concentration (MAC): See **threshold limiting value**.

MeV: Million electron volts. An electron volt is the energy associated with the movement of a particle carrying a charge numerically equal to that of an electron through a potential difference of one volt. One MeV corresponds to 1.6022×10^{-13} joule.

Mist: **Aerosol** of liquid droplets formed during condensation or from an atomizing, splashing, or foaming liquid.

Net positive suction head (NPSH): A parameter used in pump characterization to predict or prevent **cavitation**. NPSH is determined from the physical properties of the liquid and the operating conditions at the pump suction. It represents the excess of the total head of the fluid above its vapor pressure at operating temperature. Each pump has its own demand for a minimum value of this parameter; this is the **required NPSH** and is defined by the pump supplier. The **available NPSH** is determined by operating temperature and pressure, elevation of the fluid above the pump suction, and the design of the piping system. Available NPSH must always be greater than required NPSH; this must be assured by proper design and installation of the system.

Nondestructive examination: A method of examination or inspection of a material or object which does not destroy or otherwise degrade the subject.

Oxidizer: A substance which can react to increase the oxidation state of another; in the context of a hazard, a substance which can react rapidly and exothermically with certain other substances.

PEL (Permissible exposure level): Defined by OSHA; closely follows the TLV–TWA established by the ACGIH.

Portable tank: Any container designed primarily to be temporarily attached to a vehicle, not including rail tank cars or marine vessels. They are usually equipped with lugs, skids, or other mountings that allow mechanical handling of the tank.

Pressure-relief valve: Pressure- or temperature-activated device used to prevent buildup of pressure beyond a predetermined maximum that could cause a system to rupture. The design and selection of the device are dictated by codes and standards governing the safe design of the vessel or system to which it is connected. Also referred to as a relief valve, safety valve, or safety relief valve.

Rad (r): Amount of radiation which results in the absorption of 100 ergs by one gram of material.

RBE: Relative biological effectiveness. Ratio of the biological effect of a given amount of radiation to that produced by the same amount of low-energy radiation (see Table 2-4).

Relief valve: See **pressure relief valve.**

Rem: Roentgen equivalent man. Used to monitor radiation hazards; calculated by the formula rem = RBE × rads. Normal background levels are 0.01–0.02 mrem/hr. See Table 2-4.

Rupture disc: A pressure-relief component mounted in the relieving device and designed to rupture at a predetermined pressure so that fluid will be discharged and the probability of rupture of the vessel or system containing the fluid reduced.

Safety valve: See **pressure-relief valve.**

Smoke: Carbonaceous particles usually less than 0.1 micron in size. The result of incomplete combustion; may contain both liquid and solid particles.

Stonewall: A term applied to choking at the discharge of a fan or a compressor during operation. Stonewall is evidenced by gas velocities approaching the speed of sound, rapid decreases in discharge pressure, and noticeable fluctuations in pressure without changes in gas flow rates.

Stress: A measure of the effect of a load applied to a surface or cross-section of a structural component. Stress has the units of force per unit area.

Stress concentration: A localized high stress in a mechanical member, produced by changes in section or irregularities in geometry. See **stress-concentration factor.**

Stress-concentration factor: Ratio of actual stress to calculated stress on an undistributed mechanical member.

Stress raiser: An irregularity which causes a local concentration of stress in a mechanical member. Examples are notches, holes, sharp edges, and threads.

Surge: A term applied to oscillations in pressure and gas flow in fans and centrifugal compressors. The condition occurs below the lower operating limit of the gas mover and is evidenced by potentially damaging vibrations and accompanied by low-frequency booming sounds or high-pitched squeals.

Threshold limit value (TLV): Limiting values established by OSHA or the ACGIH from toxic response vs. dosage data. Formerly referred to as maximum allowable concentration (MAC). There are three types of TLV:

TLV-C Ceiling limit. The concentration which should never be exceeded, even instantaneously. When continuous monitoring is not feasible, this may be expressed as a 15-min STEL.

TLV-STEL Short-term exposure limit. The maximum time-weighted concentration of a toxic agent to which a worker can be exposed for a period not exceeding fifteen minutes without experiencing irreversible tissue damage, impaired self-rescue capability, reduced efficiency, or intolerable irritation. These periods are limited to a maximum of four per day with a minimum of sixty minutes between episodes. OSHA may specify STEL to another time limit; e.g., STEL (30).

TLV-TWA   Time-weighted average limit based on a normal 8-hour work-day or 40-hour workweek to which most workers can be exposed without adverse effects. Excursions above this limit but within TLV-C and TLV-STEL are allowed provided that they are compensated for by excursions below TLV-TWA during the period examined.

Toxic dose: The dosage at which the effects of a material are noticeable and irreversible. Data are often plotted on response vs. dosage curves or listed as, for example, TD_{10} or TD_{50}. TD_{50} is the dose at which symptoms are toxic for 50% of the reported subjects. Also see **effective dose** and **lethal dose**.

Toxicity: The property of a physical or chemical agent that produces an adverse effect on a biological organism.

Upper explosive limit (UEL): See **explosive limits**.

Upper flammability limit (UFL): See **flammability limits**.

Vapor: A gaseous phase which can be condensed to a liquid or a solid at normal pressures.

1

INTRODUCTION

1.1. OBJECTIVE

The objective of this volume is to provide, under one cover, guidance in performing many basic operations encountered in chemical facilities in a safe and professional manner. The book endeavors to achieve the following:

- Provide a comprehensive reference for certain fundamental operations.
- Provide guidance in applying appropriate practices to prevent accidents.
- Indicate unsafe and inadequate practices that should be avoided.

The audience visualized in the preparation of this book is the personnel of an independent chemical operation located in the United States that may lack a full complement of specialists. Such plants may tend to be below median size and to emphasize batch processing. The equipment selected for specific discussion reflects this orientation.

1.2. SCOPE

The book selectively discusses operations not extensively covered in readily available publications. These operations are usually covered by proprietary safety standards and operating procedures which are not generally accessible to the public. The book cannot possibly include every conceivable safety fundamental and therefore focuses on the more obvious and more frequently occurring operations.

The book emphasizes general principles and the fundamentals of organizing a safety program. From the vast literature on this subject, we have chosen to emphasize the material most widely used and most readily available to an operation without complete library facilities. Accordingly, there are many references to recognized standards and published US government regulations. The reader is assumed to have a general familiarity with these sources, but one section of the book lists relevant agencies and societies along with their addresses. Many of the primary references are to material published in the *Federal Register* (FR) or the *Code of Federal Regulations* (CFR). Typically, these are referred to by their location in the CFR in the abbreviated form (title number)-"CFR"-(section number), for example, 29CFR1910.

1.3. ORGANIZATION

Chapter 2, Materials/Chemicals Handling, discusses handling of chemical substances and materials. Safe handling requires knowledge of the material's hazardous properties and conditions created during storage, transport, and processing, and one section covers general material hazards. Liquids, solids, gases, and waste materials then are covered separately.

Chapter 3, Process Equipment and Procedures, covers safety considerations in the design, installation, operation, and maintenance of selected process equipment and auxiliary systems.

Design considerations are those which pertain to the various types of equipment covered in the chapter. Safety in design should begin at the earliest stages by considering a facility as a whole and striving for inherent safety [1].

Chapter 4, General Topics, deals with safety topics not directly related to specific operations. Principal subjects include the inspection and maintenance of equipment, spare parts handling, storage and warehousing, plant modifications, hazardous work, and worker protection.

Chapter 5, Cleanup and Process Changeover, identifies potential problems associated with safe and efficient cleanup operations. It addresses process planning, the cleaning process, the problem of changeover, and methods of equipment preparation.

Chapter 6, Training, stresses the need for educating operating personnel. Applying appropriate methods in the performance of their jobs is a key component in promoting plant safety.

Chapter 7, Plant Safety Programs and Auxiliary Topics, covers plantwide systems and more general programs whose purpose is the improvement of the overall plant safety performance.

A glossary and a compilation of abbreviations and acronyms are also provided.

REFERENCES

1. Englund, S. M., Opportunities in the Design of Inherently Safer Chemical Plants. In *Advances in Chemical Engineering*, Vol. 15, pp. 73–135, Academic Press, San Diego, 1990.

2

MATERIALS/CHEMICALS HANDLING

This chapter considers the nature and mitigation of the hazards associated with the movement and handling of materials. Later chapters will deal with specific equipment and plant procedures.

First, the chapter considers standard methods for classifying material hazards (2.1). Since safe handing requires knowledge of associated hazards (e.g., static electricity), these are dealt with next (2.2). The discussion then turns to the specifics of handling materials, in the sequence liquids (2.4), solids (2.5), gases (2.6), and wastes (2.7).

In this discussion, the term "material" refers to both chemical and non-chemical substances. "Chemicals" are those substances with readily definable compositions, such as calcium chloride, sulfuric acid, vinyl acetate, and methyl isocyanate. Nonchemicals are those substances whose chemical structures are not readily defined, such as wood, hydraulic oils, cereal grains and ceramics.

The safe handling of a given material in any situation requires consideration of two factors:

- intrinsically hazardous properties of the material
- hazardous conditions created during storage, transport, and processing

These properties can be found in standard references [1, 2, 3] and in the Material Safety Data Sheets (MSDS) required for all hazardous substances and available from suppliers. Figure 2-1 shows an MSDS form with the information requirements contained in OSHA 29 CFR 1910.1200(g). Many companies use the more comprehensive ANSI Standard Z400.1, Preparation of Material Safety Data sheets, which may run 8–12 pages in length. If the material is a specialty product manufactured on site, the properties must be determined and documented by the plant owner.

2.1. HAZARDOUS PROPERTY IDENTIFICATION

Since the enactment of the OSHA Hazard Communication (HAZCOM) regulation [4, 5], it is the responsibility of each processing facility to communicate

Material Safety Data Sheet	U.S. Department of Labor	
May be used to comply with OSHA's Hazard Communication Standard. 29 CFR 1910.1200. Standard must be consulted for specific requirements.	Occupational Safety and Health Administration (Non-Mandatory Form) Form Approved OMB No. 1218-0072	

IDENTITY (As Used on Label and List)	Note: Blank spaces are not permitted. If any item is not applicable, or no information is available, the space must be marked to indicate that.

Section I

Manufacturer's Name	Emergency Telephone Number
Address (Number, Street, City, State, and ZIP Code)	Telephone Number for Information
	Date Prepared
	Signature of Preparer (optional)

Section II — Hazardous Ingredients/Identity Information

Hazardous Components (Specific Chemical Identity; Common Name(s))	OSHA PEL	ACGIH TLV	Other Limits Recommended	% (optional)

Section III — Physical/Chemical Characteristics

Boiling Point		Specific Gravity (H_2O = 1)	
Vapor Pressure (mm Hg.)		Melting Point	
Vapor Density (AIR = 1)		Evaporation Rate (Butyl Acetate = 1)	
Solubility in Water			
Appearance and Odor			

Section IV — Fire and Explosion Hazard Data

Flash Point (Method Used)	Flammable Limits	LEL	UEL
Extinguishing Media			
Special Fire Fighting Procedures			
Unusual Fire and Explosion Hazards			

(Reproduce locally)   OSHA 174, Sept. 1985

FIGURE 2-1. Material Safety Data Sheet.

Section V — Reactivity Data

Stability	Unstable		Conditions to Avoid
	Stable		

Incompatibility (Materials to Avoid)

Hazardous Decomposition or Byproducts

Hazardous Polymerization	May Occur		Conditions to Avoid
	Will Not Occur		

Section VI — Health Hazard Data

Route(s) of Entry:	Inhalation?	Skin?	Ingestion?

Health Hazards (Acute and Chronic)

Carcinogenicity:	NTP?	IARC Monographs?	OSHA Regulated?

Signs and Symptoms of Exposure

Medical Conditions
Generally Aggravated by Exposure

Emergency and First Aid Procedures

Section VII — Precautions for Safe Handling and Use

Steps to Be Taken in Case Material Is Released or Spilled

Waste Disposal Method

Precautions to Be Taken in Handling and Storing

Other Precautions

Section VIII — Control Measures

Respiratory Protection (Specify Type)

Ventilation	Local Exhaust		Special
	Mechanical (General)		Other

Protective Gloves		Eye Protection

Other Protective Clothing or Equipment

Work/Hygienic Practices

U.S.G.P.O: 1986-491-529/4

FIGURE 2-1. Material Safety Data Sheet (continued).

to its personnel the hazards of all materials on the site. The process of identifying material hazards in the plant should include the following:

- list of all materials received, stored, used, or manufactured on site
  —raw materials
  —intermediates
  —finished products
  —maintenance materials
  —laboratory reagents and supplies
  —social facility supplies (offices, cafeteria, dispensary)
- MSDS for all materials in the above list
- all storage vessels affixed with the appropriate NFPA Diamond consistent with the 704M system [6]
- DOT shipping label [7] on all materials shipped or dispatched

Typical DOT labels are shown in Figure 2-2. Another labeling system has been devised by the NFPA in its Standard 704 for the identification of hazards. The symbol used is a diamond divided into its four corners. Each of the corners represents a different type of hazard. Health, fire, and reactivity hazards are quantified to a degree by a numeral in the appropriate field. These labels allow recognition of hazards at a glance and are widely used on containers, storage tanks, items of operating equipment, and certain area of buildings. Figure 2-3 explains NFPA's 704M Diamond System, and Figure 2-4 shows diamond designations for typical process materials.

The name of the material or one of the following identification numbers should be placed on or under the diamond:

- Chemical Abstracts Service (CAS) registry number
- NIOSH Registry of Toxic Effects of Chemical Substances (RTECS) number
- Department of Transportation identification (DOT ID) number [8]

Identification numbers for the hazardous substances included in Figure 2-4 are in Table 2-1.

Toxic substances are regulated under CFR Title 40. Before a substance can be manufactured, it must be properly registered with the EPA in accordance with the Toxic Substance Control Act (TSCA). In the case of a new substance not on the TSCA inventory, a premanufacturing notice (PMN) must be submitted to the EPA. This notice must contain information on:

- identity
- use
- anticipated production volume
- hazards
- disposal characteristics of the substance.

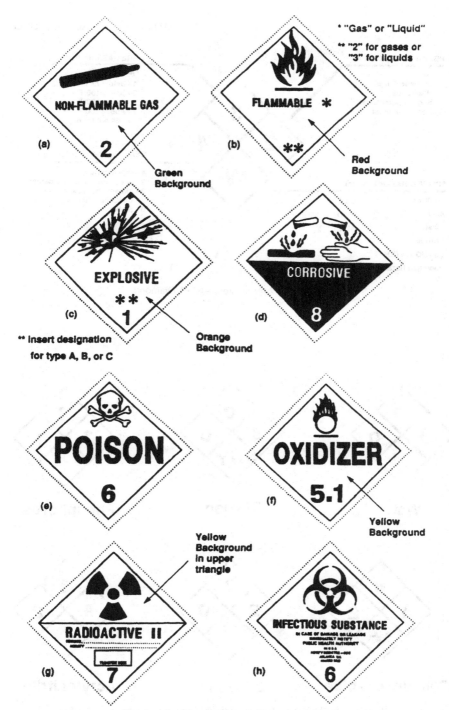

FIGURE 2-2. Examples of DOT Transportation Placards. (a) Nonflammable compressed gases; (b) flammable materials; (c) explosives; (d) corrosive materials; (e)poisonous material (for gases subsitiute POISON GAS and numeral 2 at base of placard; (f) oxidizers [5.1], organic peroxides [5.2], and oxygen [2]; (g) radioactive materials (type I, II, or III must be indicated); (h)hazardous biologicals.

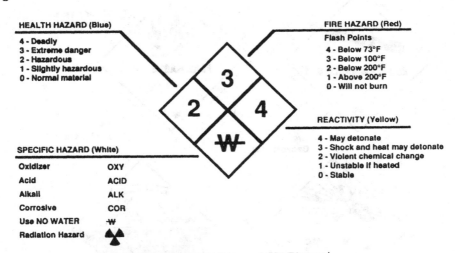

FIGURE 2-3. 704M System—NFPA Diamond.

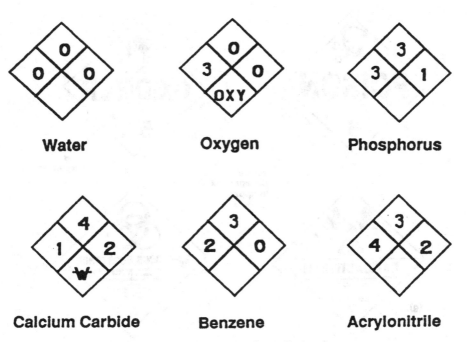

FIGURE 2-4. Typical 704M System Designations.

TABLE 2-1 Material Identification Numbers			
Material	CAS Registry No.	NIOSH RTECS Number	DOT ID Number
Acrylonitrile	107-13-1	AT5250000	1093
Benzene	71-43-2	CY1400000	1114
Calcium Carbide	75-20-7	EV9400000	1402
Oxygen	7782-44-7	RS2060000	1072 or 1073
Phosphorus	7723-14-0	TH3500000	1338, 1381, or 2447

It is the seller's responsibility to see that TSCA requirements are met and that any changes in process, formulation, etc. that change the conditions of registry are reported. The seller is also obliged to include an MSDS with each shipment. The importer of a new substance has the same responsibilities under TSCA.

It is wise to consult a number of sources of hazard information and to include the results in training programs with feedback (see Chapter 6). Surveys show that workers typically understand about two-thirds of the information on an MSDS, and OSHA claims that 90% of MSDS do not contain accurate information in all required categories [9].

2.2. MATERIAL HANDLING HAZARDS

This section addresses the hazards associated with in-plant handling of materials. After treating fire and explosion hazards and discussing chemical toxicity, it briefly discusses biological and radiation hazards. Electrical hazards are treated next, with special attention to the sources and control of static electricity. Finally, the section mentions the thermal hazards (hot and cold) associated with material handling. The emphasis in this chapter is on the material transport process. The safe operation of equipment and systems is the subject of Chapter 3.

2.2.1. Fire and Explosive Properties

Fires and explosions are the large-scale accidents most to be feared in the typical process plant. For thorough coverage of the vital topic of protection from these hazards, the reader should consult the large body of specialized literature. The National Fire Protection Association (NFPA), the publisher of the National Fire Codes, provides good summaries of codes and safe practices.

Some of the NFPA Standards consulted during preparation of this volume are listed in the Bibliography. A consolidated source is the NFPA's *Flammable and Combustible Liquids Code Handbook*. A more specialized discussion of building exits and fire escapes is in the *Life Safety Code Handbook*.

Fires and explosions can result from the rapid reaction of a material exposed to an oxidizing agent. Depending on the speed at which the reaction occurs, it is classified as a fire (relatively slow) or an explosion (rapid). In order for a fire to occur three elements must be present: fuel, oxygen, and a source of ignition. Remove any one item from the triad and combustion cannot result. Safety is improved by identifying situations that allow these three elements to come together and taking steps to prevent these situations.

The fuel element of the fire/explosion triad can be any flammable material. Degree of flammability is indicated by a material's explosive or flammable limits, its flash point, and its autoignition temperature. These, with other combustion properties, are documented in the standard references previously noted.

Oxygen for combustion is present in air, but other chemical oxidizers such as halogens, nitric acid, nitrogen oxides, and peroxides may be present. Certain materials (e.g., ethylene oxide) contain internal oxygen which increases sensitivity and can take part in the combustion reaction.

The source of ignition can be an open flame, a spark, a hot surface, static electricity, or autoignition. Plant personnel must strive to eliminate all ignition sources from locations where flammable materials are stored or handled or must institute procedures to control the resulting hazards.

Explosions or hazardous releases can also be caused by interactions between materials improperly brought together. Such hazards can be identified by establishing a matrix of all materials present in a plant or an area and identifying the possible interactions. This effort can be helped by the EPA's REACT Program, which determines likely consequences by considering the reactivities of chemical groups.

2.2.1.1. Flammable Mixtures

When a hazardous material may be at an explosive concentration or temperature during normal or upset conditions, the hazard can be lessened by:

- reduction of oxygen concentration by purging with an inert gas (inertion)
- reducing the amount of flammable material in the system or diluting it with an inert to a concentration outside the flammable range
- installing flame arresters or other isolation devices at proper locations to prevent propagation of a flame front
- changing parameters that affect autoignition
  —flow conditions
  —vapor volume or concentration
  —system pressure
  —presence and quantity of possible catalysts

The concentration of the oxidizing agent in the mixture may be increased by:

- leakage into the system through faulty joints or fittings or poorly maintained equipment, especially when under vacuum
- accidental addition by operator error

This condition can be eliminated or mitigated by:

- improved sealing
- eliminating joints and fittings or replacing them with types less susceptible to leaking
- more thorough periodic maintenance of equipment
- arranging the system and its controls to reduce the potential for operator error
- personnel training in safe operating procedures
- oxygen monitors and alarms

2.2.1.2. Ignition Sources

Plant design and procedures should be examined for potential ignition sources. Table 2-2 shows the results of an investigation of over 25,000 fires [10].

TABLE 2-2 Ignition Causes Identified by Factory Mutual Investigation	
Electrical (wiring of motors)	23%
Smoking	18%
Friction (bearing or broken parts)	10%
Overheated materials (abnormally high temperatures)	8%
Hot surfaces (heat from boilers, lamps, etc.)	7%
Burner flames (improper use of torches)	7%
Combustion sparks (sparks and embers)	5%
Spontaneous ignition (rubbish, etc.)	4%
Cutting and welding (sparks, arcs, heat, etc.)	4%
Exposure (fires jumping into new areas)	3%
Incendiarism (fires maliciously set)	3%
Mechanical sparks (grinders, crushers, etc.)	2%
Molten substances (hot spills)	2%
Chemical action (processes not in control)	1%
Static sparks (release of accumulated energy)	1%
Lightning (where lightning rods are not used)	1%
Miscellaneous	1%

Ignition sources also result from improper installation of electrical equipment. Besides observing safe operating practices, the process plant operator should verify that electrical systems are properly installed and enclosed or are intrinsically safe. Such installation should always be in accordance with NEC codes and NEMA standards [11, 12]. Instrument systems are covered in ISA standards and are discussed in Section 3.12. Other sources include:

- flares
- furnaces
- engines
- motor vehicles
- welding and other hot work
- static electricity accumulation.

2.2.2. Chemical Toxicity

"Toxicity" relates to the adverse effects of substances on organisms. Effects may vary with the nature of the chemical compound and the extent of exposure. Specific information on toxicity can be found in a variety of sources [1, 2, 3, 13, 14]. The toxicity data on the MSDS must include the following:

- list of hazardous components
- threshold limit values: TLV-TWA, TLV-STEL, TLV-C, or PEL
- mechanism by which exposure occurs
  —ingestion
  —inhalation
  —dermal absorption
- health hazards
  —acute
  —chronic
- symptoms of exposure

Table 2-3 lists relative degrees of toxicity [15]. The criterion is the size of dose that is lethal to 50% of a population.

Once a material is identified as toxic, the nature of its toxicity should be documented and recorded on an MSDS. Standardized descriptions of the symptoms should be used as much as possible. A comprehensive list of MSDS designations with a description of associated symptoms can be found in [16].

Toxic exposure can occur from

- malfunctioning equipment and safety systems
- failure of personal protective equipment
- unplanned entrance of personnel into hazardous areas
- toxic materials that are normally foreign to the plant or are by-products of abnormal operation
- improper sampling or handling of samples

TABLE 2-3 Toxicity Rating Table				
		Experimental LD$_{50}$ for Rats		
Toxicity Rating	Degree of Toxicity	Single oral dose per kg body weight	4 hr. Inhalation dose ppm	Probable LD$_{50}$ for a 150 lb person
1	Extremely toxic	<1.0 mg	<10	A taste
2	Highly toxic	1.0–50 mg	10–100	A teaspoon
3	Moderately toxic	50–500 mg	100–1000	An ounce
4	Slightly toxic	0.5–5 g	1,000–10,000	A pint
5	Practically nontoxic	5–15 g	10,000–100,000	A quart
6	Relatively harmless	>15 g	>100,000	More than a quart

Proper design and maintenance will reduce the probability of system failure. Operator training and the use of barriers and visible warnings will reduce the number of the human errors. When toxic materials are handled or are routinely present in an area, the following precautions therefore are appropriate:

- The area is posted, and access is restricted.
- Detailed written procedures are available to personnel and are carefully followed.
- Only qualified personnel who have had prescribed training are permitted to work in the area.
- Emergency escape equipment is available, and evacuation routes are established.
- A permit system (see Section 7.1.4) is in place, and permits are required for certain maintenance and nonroutine work.
- Work areas are well ventilated, and the atmosphere in an enclosed space is tested periodically or before entry.
- Exposure time is limited by rotating personnel in and out of the area.
- Ambient air monitors or gas analyzers are used to detect the presence of hazardous materials.

2.2.3. Biological Hazards

The production of pharmaceuticals, the manufacture of enzymes or genetically engineered materials, and the cultivation of biological materials can expose personnel to biologically produced toxic agents. Some of the situations that may create biological hazards are familiar to chemical operators:

- poor process layout
- inadequate or poorly maintained equipment
- inadequate training of personnel in handling techniques

Some of the methods of defining toxic dosages, mechanisms of toxic agent transport, and symptoms for chemical toxicity referenced in Section 2.2.2 can also be applied to biologicals. DHHS has a method for establishing biosafety levels and has published a general operating guide [17].

2.2.4. Radiation Hazards

Many processes and inspection procedures depend on radiation-emitting substances and equipment. Potential sources of exposure include:
- radiographic examination of equipment
- smoke detectors using alpha-emitting radioisotopes
- radioisotopes used as tracers in fluid flow and biochemical analysis
- radiation-based level and density instruments
- microwave ovens
- industrial lasers

TABLE 2-4 Parameters Used to Define and Monitor Radiation Hazards	
Relative Biological Effectiveness (RBE) of Common Types of Radiation:	
Radiation Type	*RBE*
Low energy beta, gamma, and X-rays, electrons (less than 0.03 MeV)	1[a]
High energy beta, gamma, and X-rays, electrons (greater than 0.03 MeV)	1.7
Thermal neutrons (low energy)	3[a]
Fast neutrons and protons to 10 MeV	10
Alpha particles[b]	10
Radiation exposure action limits:	
Exposure Level	*millirem/hr*
Normal background	0.01–0.02
Above background. Suggests caution and investigation of source of radiation. Total exposure is limited to 1–5 remannually.	0.02–0.2
Serious radiation hazard. Evacuate site. Continue investigation following the advice of an established health physicist who is knowledgeable in radiation safety.	≥2
[a]Typical radiation in inspection and detection devices. [b]Typically found in smoke detectors. Alpha particles pose a particular hazard through inhalation and ingestion.	

Table 2-4 lists the parameters used to monitor radiation as well as some of the exposure limits. Definitions of terms are in the Glossary. Recommended practice for fire protection is in NFPA 801.

Some general guidelines for handling sources of radiation safely are:

- Provide adequate labeling of radioactive materials (29CFR 1910.1200(f)).
- Prepare and implement employee emergency plans and fire prevention plans (29CFR 1910.38). In case of accident:
  —Notify employees, emergency personnel, the community, and governmental licensing agencies.
  —Identify the nature of the hazard.
  —Identify the limits of the area affected by the accident.
  —Reduce the duration of personnel exposure.
- Provide adequate shielding around the source to minimize personnel exposure. Monitor shielding for damage, degradation, and radiation leaks.
- When using radiation equipment, make sure that:
  —the equipment is operable
  —adequate operating and handling instructions are provided
  —trained, qualified personnel are available to operate and maintain the equipment
- Provide adequate storage that reduces personnel exposure and keeps radioactive materials away from potentially hazardous locations (e.g., areas storing or handling combustible materials).

2.2.5. Electrical Hazards

Four categories of electrical hazard are discussed below:

- hazards related to electrical equipment and wiring
- hazards from static electricity generation and discharge
- hazards from lightning
- hazards from stray currents

Electricity is an ignition source that can cause fires and explosions if equipment is installed or wired inadequately or if a static discharge occurs. Table 2-2 showed that electrical systems are the leading cause of fires. Improperly installed electrical systems may also cause accidental electrocution. Stray currents may result in corrosion of piping and equipment.

2.2.5.1. Electrical Equipment and Wiring Hazards
Installation of Equipment
Installation of electrical equipment is governed by various codes and standards. Compliance with the National Electrical Code (NEC), NFPA 70, is required by both OSHA and insurance companies. The plant owner must also

comply with various state and local regulations for electrical installation. Some of the factors to consider are:

- The hazardous nature of the area in which the equipment is installed. Table 3-8 outlines the NEC method for determining the Group, Class, and Division of a hazardous area.
- The enclosure provided for the equipment. Table 3-10 lists the NEMA enclosures recommended for various environments.
- Motor enclosures. Table 3-9 lists the various NEMA enclosures for electric motors.
- Purging of electrical and instrumentation cabinets to prevent accumulation of combustible gases and vapors.
- Use of intrinsically safe systems.
- Short circuit and overload protection.
- Proper maintenance and grounding.

Grounding of Equipment and Wiring

A properly installed grounding system protects against ground faults, static discharge, and lightning. The resistance, and so the effectiveness, of the ground path is affected by five factors, almost all of which can be controlled:

- surface condition of the grounded element
- installation of the grounding and bonding elements
- installation of the grounding and bonding connectors
- material and size of the conductor and electrode
- nature of the soil

The relationships among these factors are shown in Figure 2-5.
Some guidelines to follow for good grounding and bonding are:

- Size grounding and bonding conductors in accordance with NEC Article 250.
- Make sure that mating surfaces of grounding and bonding connections are clean, smooth, and free of corrosion, paint, and other insulating substances.
- Keep grounding and bonding terminals accessible.
- Use corrosion-resistant hardware for grounding and bonding connections.
- Clearly identify grounding terminals with appropriate color coding or other standard method.
- Protect grounding components from physical damage.
- Make sure that all connections are properly tightened.
- Upgrade the grounding system as required to accommodate load increases or equipment relocation.

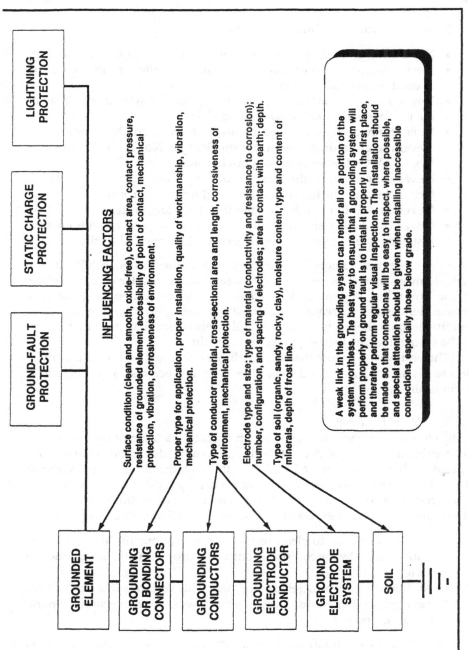

FIGURE 2-5. Factors Influencing Integrity of Grounding Systems.

Some prohibitions to observe are as follows:

- Never use a grounding terminal for any purpose other than grounding.
- Never connect any portion of the grounding system so that it can be opened by a switch or circuit breaker.
- Never use the grounded conductor of a grounded system other than in the manner permitted by NEC Article 250.
- Never connect series overcurrent devices to the grounded connections.
- Never use grounding conductors smaller than No. 14 AWG copper or No. 12 AWG aluminum or copper-clad aluminum.
- Never connect more than one grounding conductor to a grounding terminal unless the terminal is designed for that purpose
- Never use lightning rods for other than lightning protection
- Never disconnect any portion of a grounding system without first installing a temporary jumper across the disconnected section

Control of these factors is relatively simple in the design, installation, or maintenance stages. However, a single weak link can render all or a portion of the grounding system useless. Detection and correction of grounding system deficiencies increases in difficulty as corrosion, vibration, weather conditions and other factors take their toll. This is especially true in portions of the system that are below grade or inaccessible.

The best way to ensure that a grounding system will perform properly is first to install it correctly and then to visually inspect, test, and maintain it periodically.

2.2.5.2. Static Electricity

Static electricity is generated by the friction and separation of two dissimilar nonconducting materials. A positive charge accumulates on one material and a negative charge on the other. Eventually, the potential difference between the two materials or between one of them and a nearby ground can allow sudden discharge in the form of a spark. Static electricity is an insidious hazard whch often appears when not expected. Two examples will illustrate this point:

- The use of steam to clean residues from equipment often produces static charges by the collision of entrained droplets with nonconductive material.
- Addition of a dry powder to equipment with a combustible atmosphere can produce dangerous charges on the powder or on ungrounded parts of the equipment.

In order for static electricity to cause a fire or explosion, several conditions must exist:

- a source of static generation
- accumulation of electric charges

- the presence of an ignitable mixture
- a spark discharge of sufficient energy

The most frequent hazards of static electricity buildup in an operating plant are as follows:

- Sparks capable of igniting hazardous materials. The minimum ignition energies for various gases and dusts are shown in Figure 2-6. A dust can create its own ignition source when charges are generated on particles which then move through plant equipment.
- Incendive sparks where low-conductivity flammable or combustible liquids discharge into containers. The liquids can mix with air to produce explosive mixtures.
- Physical contact by personnel with electrostatically charged objects. The resulting shocks can cause uncontrolled reflex actions.
- Difficulties or accidents due to statically charged materials, such as the use of belt drives in dusty atmospheres, particularly if the belts cross and rub continuously.

Methods of Reducing the Hazards of Static Electricity

Several techniques can be used to reduce the probability of static discharges:

- grounding and bonding
- maintaining high atmospheric humidity
- increasing the conductivity of air by ionization
- use of conductive materials where practicable
- increasing the conductivity of nonconductors with additives
- reduction of velocities of fluids in pipelines
- use of nonmetallic shields or guards to prevent contact of personnel with metal parts
- avoiding transfer of nonconductive materials through nonconductive equipment, pipelines, and containers
- avoiding use of nonconductive containers wherever practicable
- avoiding transfer of nonconductive materials through nonconductive atmospheres (e.g., free fall of material through dry air)

Grounding and Bonding for Static Electricity Hazards

Grounding and bonding prevent the accumulation of static electricity by draining charges to ground before a dangerous potential develops. Current flow during static discharge is small, on the order of a microampere. Potential differences may be thousands of volts. A ground or bond resistance as high as a megohm may therefore still allow static dissipation. Methods for attaining high-quality grounding and bonding are discussed in Section 2.2.5.1.

Specific methods that apply to material handling operations are:

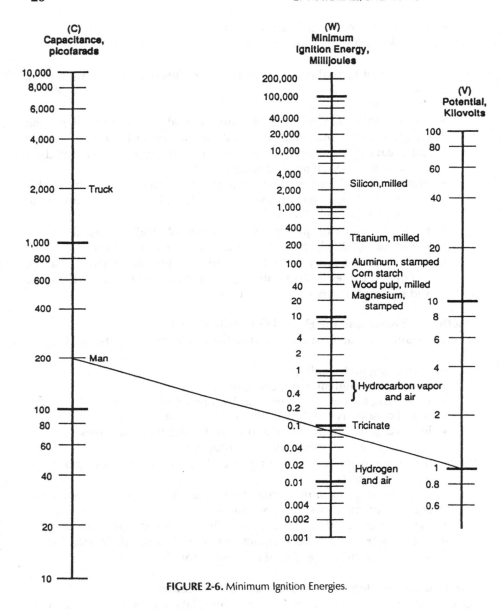

FIGURE 2-6. Minimum Ignition Energies.

- When flammable or combustible liquids are transferred from drums or tanks into metal containers, the vessels should be bonded together and connected to the nearest ground. Transfer hoses should be bonded to the grounded vessel. Figure 2-11 in Section 2.4.1 shows one example.
- Storage and handling equipment not firmly connected to a grounded steel structure should be individually grounded.

- Pipelines carrying flammable or combustible liquids should be grounded at points of entry to classified hazardous areas.
- When nonconductive containers are to be filled with flammable liquids, charges can be dissipated by bottom filling with a grounded lance. The person filling the container should be provided with a grounding connection and instructed to avoid rubbing the external surface of the container.
- In a piping system, valve handles should be bonded to the valve bodies or to a metallic pipe to prevent static buildup on the handles.
- When cargo tanks containing flammable vapors are loaded through top domes, they should be bonded to the fill stem, piping, or steel loading rack. If bonding is to the rack, it is essential that the piping and rack be electrically interconnected. The bond connection should be made before the dome cover is open and should remain in place until loading is complete and the dome cover securely closed. Bottom or dip pipe filling to eliminate the free fall of liquid reduces the hazard of static accumulation.
- When solid nonconductive materials are processed, all metal parts of the process equipment should be bonded and grounded.
- Workers cleaning the inside of a tank should be grounded directly with wrist bands and have proper footwear.

Humidification

The conductivity of certain insulators such as paper and plastics depends on their moisture content. A higher moisture level facilitates dissipation of static electricity.

A high humidity in the surrounding air (>60% RH) often prevents the accumulation of static charges. A microscopic film of moisture covers the surfaces, making them more conductive and decreasing the hazard of static electricity. This method sometimes is used in areas with explosive dust hazards, but it is often difficult to apply and of doubtful value.

Another method with similar effects is the use of anti-static materials and additives to increase the rate of dissipation of electrostatic charges. Nitrogen compounds, sulfonic acids, polyglycols, and polyhydric alcohols are examples of anti-static materials. They can be sprayed or wiped on surfaces or added to bulk materials.

Ionization

Ions are produced in air by high temperatures or ultraviolet radiation. The ionized air can be sufficiently conductive to reduce static hazards. The static charge is conveyed by the ionized air from an object to a grounding electrode located nearby. When UV lamps are used, eye protection is required.

Nonmetallic Guards and Shields

Additionally, workers can be physically separated from electrostatically charged materials such as metal handrails or door knobs. Nonmetallic shields or guards will prevent direct contact of workers with metal parts.

2.2.5.3. Lightning Hazards

Lightning is a static discharge between the earth and a storm cloud. While external to a process, it is an important problem because of its intensity. Lightning protection reduces hazards to life and equipment and helps to prevent electrical service interruptions. The choice and design of protective devices depend on:

- degree of exposure to the hazard (frequency and intensity of thunderstorms; heights of stacks and other structures)
- required level of protection (inherent fire risk; consequences of a lightning strike)
- economic risk (higher-cost protection should be considered in larger plants, continuous operations, situations requiring increased safety, etc.)

Protective systems can use special devices or connect to existing grounding systems.

- Protective devices
    —For buildings and structures, use rods and masts. See Figure 2-7 for the cone of protection offered by a mast. Masts must be grounded and interconnected with the grounding system of the structure.
    —For distribution lines, use overhead static wires connected to ground.
    —For electrical equipment, use surge arrestors and capacitors.
- Grounding and bonding
    —Use ground electrodes even with equipment and structures that are inherently grounded.
    —Give special attention to
        –masonry stacks and structures
        –vessels containing flammable materials
        –rotating equipment
        –outdoor piping
        –conveyors, elevators, etc.
    —Follow NEC recommendations.
    —Use short, direct paths to ground while avoiding contact with combustible atmospheres.

The following are typical grounding procedures for important plant equipment and structures:

- Metal buildings and other metal structures should be connected at the base to one or more ground electrodes by two or more independent connections (see Figure 2-8). Building structural steel can be used as the main conductor of the lightning protection system if it is inherently continuous or bonded together to provide electrical continuity.
- The shells of all tanks in hazardous locations should be effectively grounded, whether or not they contain combustible substances. A tank

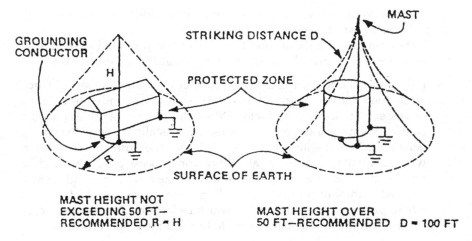

FIGURE 2-7. Ligntning Rod Protection. A lightning rod provides a cone of protection for a;ll structures within the cone. When a rod is more than 50 feet high, the protected zone is defined by an arc that varies with the intensity of the discharge.

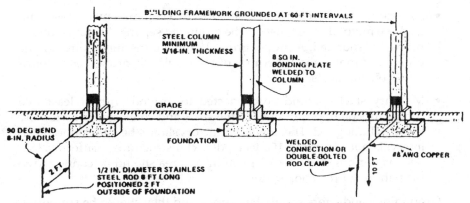

FIGURE 2-8. Grounding of Structural Steel. Building structural steel should be bonded together, with connections to ground at appropriate intervals. Connections to columns should be made above grade to permit inspection. Visual inspection is usually the most reliable; ground impedance tests can indicate a proper ground when only one strand of the ground conductor is intact.

in direct contact with the earth is considered inherently grounded if it meets all these conditions:

—The metal roof has a minimum thickness of $\frac{3}{16}$ inch.

—The roof is welded, bolted, or riveted to the shell.

—All joints between metallic plates are riveted, bolted, or welded.

—All entering pipes are connected to the tank so that there is solid, substantial metallic contact between pipe and tank.

—All uncontrolled vapor or gas openings are closed or provided with flame arresting devices.

Tanks not meeting these conditions should be provided with at least two grounding connections to separate grounding conductors.

- Piping that is not in good electrical contact with a tank or vessel (such as an open line discharging into a tank) should be bonded to the tank or vessel by a flexible conductor. Where couplings are not electrically continuous, a separate conductor should be installed to bond the piping between where it enters and where it leaves the fitting. Unless otherwise adequately grounded, all utility and process piping should be bonded to a common conductor and connected to the grounding system. At a minimum, such grounding connections should be made where piping enters or leaves a hazardous location. No pipelines containing flammable substances should be used for ground continuity.

- Each vessel should be grounded by a conductor affixed to a lug welded to the vessel. The grounding lug should be at least 18 inches above earth level to reduce the possibility of corrosion.

- An electrically driven pump assembled to a common metal base with its drive motor does not need to be grounded separately if the motor is properly grounded. A proper ground cannot be assumed through the pump and motor shafts because of the insulating properties of lubricating oil and grease.

- Masonry stacks should be connected to ground with at least two conductors on opposite sides of the stack and connected individually to the lightning rod. The down conductors should be interconnected at intervals not exceeding 100 feet. All metal structures, platforms, ladders, water pipes, etc. located within the area should be connected to the lightning protection system.

Grounding conductors should be copper, and they should be supported at intervals of not more than two to three feet. Mechanical protection should be provided for all ground system conductors that are less than eight feet above grade.

Many lightning system components gradually lose effectiveness because of corrosion, weather damage, or lightning damage. Integrity of the grounding system therefore requires special attention. A maintenance schedule should be established for all grounding systems, including rods and masts. The program should include

- visual inspection
- continuity tests
- resistance checks at all connections
- replacement of damaged components.

2.2.5.4. Stray and Galvanic Current Hazards

Stray and galvanic currents may cause:

- corrosion and eventual destruction of a piping system, a tank, or another metallic object
- fires or explosions from arcs that can ignite flammable liquids, gases, or dusts

Stray Currents

Stray currents may be underground electric currents that travel through the soil and enter metal structures. Direct currents (DC) are the most common and most damaging type. These are generated from grounded DC electric power operations including welding machines, rectifiers, electroplating processes, battery- charging equipment, and motor–generator sets. Damage by alternating currents (AC) is much less than by DC (about 1–2% of DC). Corrosion is usually greater at low frequencies.

Corrosion occurs at the anodic (low-voltage) point of the structure. Stray currents can also be present in piping and process structures. They are an inherent part of electrochemical processing, where they are usually controlled by insertion of insulators and interruption of continuity of electrolyte flow (current breakers).

Galvanic Currents

Galvanic currents occur whenever two dissimilar metals are connected by a conductor and exposed to an electrolyte. For example, if a buried copper ground rod (cathode) is connected with a wire to an underground steel piping system (anode), the electrolytic nature of the soil allows a galvanic current to flow between the metals. The current will flow through the soil from anode to cathode, and anodic metal will be dissolved.

Galvanic currents also can flow between various parts of the same metallic structure because of differences in the surrounding electrolyte or the metal itself (impurities, welding points, etc.). Metal composition, environmental factors, and the physical and chemical properties of the soil determine which areas on a metal surface become anodes and cathodes.

In a chemical plant, typical metallic objects that may be affected by stray and galvanic currents are:

- pipelines
- tanks
- armored, metal-clad, and metal-sheathed cables
- electrical conduit
- grounding system grid

Cathodic Protection

Corrosion may be prevented or reduced by cathodic protection. Since corrosion of a metal occurs at the point at which stray or galvanic direct current flows out from the material, the remedy is to reverse the direction by means of a forced direct current into the material which, at that point, becomes a cathode. This approach requires a power source that impresses a current through an anode and into the object being protected. A more passive approach is to provide electrical connections from the object to a sacrificial anode constructed of a less noble material (see Table 3-2).

2.2.6. Thermal Hazards

Some hazards arise when process temperatures are significantly different from ambient. Hot or cold surfaces or materials in process are common sources of injury. The frequency of these injuries can be reduced by promoting awareness of the hazard, mitigating the hazard, and protecting the worker from the hazard.

Thermal hazards, perhaps because they are so familiar in everyday life, sometimes are paid less attention than other hazards. Constant awareness can be promoted by training, safety meetings, and signposting. The hazards can be mitigated by enclosure of the hot or cold equipment, insulation, and, where appropriate, restriction of access to the vicinity of the hazard. Workers are protected by the use of safety gear (Section 4.7) and the availability and prompt use of simple first aid.

The working environment may also be too hot, humid, or cold for comfort. This is the subject of Section 4.8.1.

2.2.7. Physical Plant Hazards

Some safety hazards associated with material handling are the direct result of the physical work itself or the layout of the plant, such as:

- unattended or improperly shored excavations
- piping hung too low or across access ways
- poorly designed or installed stairways and landings
- poorly supported working surfaces and flooring
- steep grades
- slippery surfaces
- sharp objects
- poor lighting
- inadequate means of egress from a hazardous area

OSHA regulations require the elimination of these hazards, and a number of books have been written on the subject [18, 19, 20, 21]. Some of these hazards are discussed in more detail in Chapters 3, 4, and 5 as they relate to specific operations.

2.3. MATERIAL TRANSPORT

DOT regulations for hazardous material transport are in Subchapter C of 49CFR. This subchapter includes:

Part 171   General information, regulations, and definitions
Part 172   Hazardous materials table, special provisions, hazard communication (HAZCOM) and emergency response information requirements
Part 173   Shippers—general requirements for shipments and packaging
Part 174   Carriage by rail
Part 175   Carriage by aircraft
Part 176   Carriage by vessel
Part 177   Carriage by public highway
Part 178   Specifications for packaging
Part 179   Specifications for tank cars
Part 180   Continuing qualification and maintenance of packagings
Subchapter D covers pipeline transport.

Handling of hazardous materials in the operating plant is governed by various codes and regulations. Many of these depend on

- State of the material—Liquid, solid or gas.
- Method of transport—piping, ducts, conveyors, portable bins and tanks, etc.
- Method of storage—tanks, hoppers, pallets, etc.

Procedures, codes, and regulations for safe handling of liquids, solids, and gases are discussed in this chapter. Specific safety requirements for equipment and plant layout are presented in Chapters 3 and 4.

2.4. LIQUID HANDLING

Safe handling of liquids requires identification of their hazardous properties, chiefly:

- toxicity
- corrosiveness
- flammability
- reactivity

This information will be on the MSDS, along with precautions for safe handling and use. A hazardous liquid shipped to the plant should have the appropriate DOT label, together with an ID number and associated safety documentation. While on site, the hazardous liquid must be transported and stored according to the appropriate codes and standards (see Tables 2-5 and 2-6).

TABLE 2-5
Codes and Standards for Liquid Storage and Handling

API 12B	Bolted Tanks for Storage of Production Liquids.
API 12D	Field Welded Tanks for Storage of Production Liquids.
API 12F	Shop Welded Tanks for Storage of Production Liquids.
API 1632	Cathodic Protection of Underground Petroleum Storage Tanks and Piping Systems
API 2000	Venting Atmospheric and Low-Pressure Storage Tanks.
API 620	Recommended Rules for the Design and Construction of Large, Welded, Low-Pressure Storage Tanks.
API 650	Welded Steel Tanks for Oil Storage.
ASME Section VIII	Code for Unfired Pressure Vessels.
ASTM D 4021	Standard Specification for Glass-Fiber Reinforced Polyester Underground Petroleum Storage Tanks.
NFPA 30	Flammable and Combustible Liquids Code.
NFPA 30B	Code for the Manufacture and Storage of Aerosol Products.
NFPA 31	Standards for the Installation of Oil Burning Equipment.
NFPA 321	Standard on Basic Classification of Flammable and Combustible Liquids.
NFPA 325M	Fire Hazard Properties of Flammable Liquids, Gases, and Volatile Solids.
NFPA 33	Standards for Spray Application Using Flammable and Combustible Materials.
NFPA 34	Standards for Dipping and Coating Processes Using Flammable or Combustible Liquids.
NFPA 35	Standards for the Manufacturing of Organic Coatings.
NFPA 36	Standards for Solvent Extraction Plants.
NFPA 37	Standards for the Installation and Use of Stationary Combustion Engines and Gas Turbines.
NFPA 385	Standards for Tank Vehicles for Flammable and Combustible Liquids.
NFPA 45	Standard on Fire Protection for Laboratories Using Chemicals.
NFPA 49	Fire Protection Handbook.
NFPA 49M	Manual of Hazardous Chemical Reactions.
NFPA 50B	Standard for Liquefied Hydrogen Systems at Consumer Sites.
NFPA 58	Standard for the Storage and Handling of Liquefied Petroleum Gases
NFPA 59A	Standards for the Production, Storage, and Handing of Liquefied Natural Gas
PEI RP-100-90	Recommended Practice for the Installation of Underground Liquid Storage Systems, Petroleum Equipment Institute, 1990.
UL 1316	Standard for Glass-Fiber Reinforced Plastic Underground Storage Tanks for Petroleum Products.
UL 142	Standard for Steel Aboveground Tanks for Flammable and Combustible Liquids
UL 58	Standard for Steel Underground Tanks for Flammable and Combustible Liquids.
UL 80	Standard for Steel Inside Tanks for Oil Burner Fuel.

TABLE 2-6
Codes and Standards Indirectly Related to Liquid Storage and Handling

ANSI B31	American National Standard Code for Pressure Piping.
API Data Sheet 2216	Ignition Risk of Hot Surfaces in Open Air.
API 2003RP	Protection Against Ignitions Arising Out of Static, Lightning, and Stray Currents.
API 2015	Cleaning Petroleum Storage Tanks.
API 2015A	A Guide for Controlling the Lead Hazard Associated with Tank Entry and Cleaning.
API 2015B	Cleaning Open Top and Covered Floating Roof Tanks.
API 2021	Guide for Fighting Fires In and Around Petroleum Storage Tanks.
ASTM D 5-86	Standard Test Method for Penetration of Bituminous Materials.
ASTM D 56-87	Standard Method of Test for Flash Point by the Tag Closed-Cup Tester.
ASTM D 86-82	Standard Method of Test for Distillation of Petroleum Products.
ASTM D 502-84	Standard Method of Test for Flash Point of Chemicals Closed-Cup Methods.
NFPA 101	Life Safety Code.
NFPA 15	Standard for Water Spray Fixed Systems for Fire Protection.
NFPA 220	Standards on Types of Building Construction.
NFPA 251	Standard Methods of Fire Tests of Building Construction and Materials.
NFPA 69	Standards on Explosion Prevention Systems.
NFPA 77	Recommended Practice on Static Electricity.
CLC-S603.1-M	Standard for Galvanic Corrosion Protection Systems for Steel Underground Tanks for Flammable and Combustible Liquids, Underwriters Laboratories of Canada.
NACE RP-02-69	Recommended Practice, Control of External Corrosion on Metallic Buried, Partially Buried, or Submerged Liquid Storage Systems, National Association of Corrosion Engineers.

NFPA 30 [22] addresses the hazards of handling and storing these liquids. Table 2-7 shows the NFPA 30 classification by properties of a liquid. Portions of NFPA 30 will be discussed in subsequent sections as they apply to hazardous liquid transport, storage, and spills.

The Dow Fire and Explosion Index shown in Figure 2-9 [23] can also be used to characterize the degree of hazard. This index combines material and process factors to assess the magnitude of a hazard and the likely extent of damage by an accident. Starting from a material factor (MF), this method derives an index by applying general (F_1) and special (F_2) process hazards factors. MF reflects the hazardous nature of the material, its quantity, and the conditions of storage. The general process hazards factor, F_1, covers such considerations as the degree of enclosure of the material, the difficulty of

TABLE 2-7
NFPA 30 Classification of Flammable and Combustible Liquids

Flammable Liquid Class	Description
IA	A liquid having a flash point below 73°F and boiling point below 100°F.
IB	A liquid having a flash point below 73°F and boiling point at or above 100°F.
IC	A liquid having a flash point at or above 73°F and below 100°F.

Combustible Liquid Class	Description
II	A liquid having a flash point at or above 100°F and below 140°F.
IIIA	A liquid having a flash point at or above 140°F and below 200°F.
IIIB	A liquid having a flash point at or above 200°F.

access, and the nature of any chemical reaction. The special process hazards factor, F_2, deals with such things as toxicity of the material, temperature and pressure of storage, the possibility of leakage, and the possibility of being within the flammable region. The calculation form in Figure 2-9 will make clear the way in which the process hazards factors are derived. The fire and explosion index results from multiplication of the factors MF, F_1, and F_2.

The Dow method can be extended to estimate the probable financial loss due to an accident. This involves the application of a series of loss control credit factors, shown on the second page of Figure 2-9.

2.4.1. Liquid Transport

Hazardous liquids are transported by a variety of methods, the choice depending on the properties of the liquid and the volume transported. Shipping containers are regulated by the DOT.

2.4.1.1. Container Transport

Small quantities of hazardous liquids transported on site can remain in their DOT-approved portable containers or can be transferred to safety cans (Figure 2-10) or portable tanks. Rules for the use of safety cans and portable tanks for flammable liquids are in NFPA 30; many of the criteria used for aboveground permanent storage vessels apply. Generally, the quantity of hazardous liquids transferred by these containers to their point of use should be kept to a minimum. The risk increases markedly with the quantity of liquid accumulated in any location.

FIRE & EXPLOSION INDEX

AREA / COUNTRY	DIVISION	LOCATION	DATE
SITE	MANUFACTURING UNIT	PROCESS UNIT	
PREPARED BY:	APPROVED BY: (Superintendent)		BUILDING
REVIEWED BY: (Management)	REVIEWED BY: (Technology Center)		REVIEWED BY: (Safety & Loss Prevention)
MATERIALS IN PROCESS UNIT			
STATE OF OPERATION ___ DESIGN ___ START UP ___ NORMAL OPERATION ___ SHUTDOWN		BASIC MATERIAL(S) FOR MATERIAL FACTOR	

MATERIAL FACTOR (See Table 1 or Appendices A or B) Note requirements when unit temperature over 140 °F (60 °C)		

1. General Process Hazards	Penalty Factor Range	Penalty Factor Used(1)
Base Factor	1.00	1.00
A. Exothermic Chemical Reactions	0.30 to 1.25	
B. Endothermic Processes	0.20 to 0.40	
C. Material Handling and Transfer	0.25 to 1.05	
D. Enclosed or Indoor Process Units	0.25 to 0.90	
E. Access	0.20 to 0.35	
F. Drainage and Spill Control _____ gal or cu.m.	0.25 to 0.50	
General Process Hazards Factor (F₁)		

2. Special Process Hazards		
Base Factor	1.00	1.00
A. Toxic Material(s)	0.20 to 0.80	
B. Sub-Atmospheric Pressure (< 500 mm Hg)	0.50	
C. Operation In or Near Flammable Range ___ Inerted ___ Not Inerted		
1. Tank Farms Storage Flammable Liquids	0.50	
2. Process Upset or Purge Failure	0.30	
3. Always in Flammable Range	0.80	
D. Dust Explosion (See Table 3)	0.25 to 2.00	
E. Pressure (See Figure 2) Operating Pressure _____ psig or kPa gauge Relief Setting _____ psig or kPa gauge		
F. Low Temperature	0.20 to 0.30	
G. Quantity of Flammable/Unstable Material: Quantity _____ lb or kg H_C = _____ BTU/lb or kcal/kg		
1. Liquids or Gases in Process (See Figure 3)		
2. Liquids or Gases in Storage (See Figure 4)		
3. Combustible Solids in Storage, Dust in Process (See Figure 5)		
H. Corrosion and Erosion	0.10 to 0.75	
I. Leakage – Joints and Packing	0.10 to 1.50	
J. Use of Fired Equipment (See Figure 6)		
K. Hot Oil Heat Exchange System (See Table 5)	0.15 to 1.15	
L. Rotating Equipment	0.50	
Special Process Hazards Factor (F₂)		
Process Unit Hazards Factor (F₁ x F₂) = F₃		
Fire and Explosion Index (F₃ x MF = F&EI)		

(1) For no penalty use 0.00.

FIGURE 2-9. Dow Fire and Explosion Index (continued on next page). [From Dow's Fire and Explosion Index Hazard Classification Guide, 7th ed., 1994. Reproduced with permission of the American Institute of Chemical Engineers.]

LOSS CONTROL CREDIT FACTORS

1. Process Control Credit Factor (C_1)

Feature	Credit Factor Range	Credit Factor Used(2)	Feature	Credit Factor Range	Credit Factor Used(2)
a. Emergency Power	0.98		f. Inert Gas	0.94 to 0.96	
b. Cooling	0.97 to 0.99		g. Operating Instructions/Procedures	0.91 to 0.99	
c. Explosion Control	0.84 to 0.98		h. Reactive Chemical Review	0.91 to 0.98	
d. Emergency Shutdown	0.96 to 0.99		i. Other Process Hazard Analysis	0.91 to 0.98	
e. Computer Control	0.93 to 0.99				

C_1 Value(3) [ ]

2. Material Isolation Credit Factor (C_2)

Feature	Credit Factor Range	Credit Factor Used(2)	Feature	Credit Factor Range	Credit Factor Used(2)
a. Remote Control Valves	0.96 to 0.98		c. Drainage	0.91 to 0.97	
b. Dump/Blowdown	0.96 to 0.98		d. Interlock	0.98	

C_2 Value(3) [ ]

3. Fire Protection Credit Factor (C_3)

Feature	Credit Factor Range	Credit Factor Used(2)	Feature	Credit Factor Range	Credit Factor Used(2)
a. Leak Detection	0.94 to 0.98		f. Water Curtains	0.97 to 0.98	
b. Structural Steel	0.95 to 0.98		g. Foam	0.92 to 0.97	
c. Fire Water Supply	0.94 to 0.97		h. Hand Extinguishers/Monitors	0.93 to 0.98	
d. Special Systems	0.91		i. Cable Protection	0.94 to 0.98	
e. Sprinkler Systems	0.74 to 0.97				

C_3 Value(3) [ ]

Loss Control Credit Factor = C_1 X C_2 X $C_{3(3)}$ = [ ] (Enter on line 7 below)

PROCESS UNIT RISK ANALYSIS SUMMARY

1.	Fire & Explosion Index (F&EI)..........................(See Front)		
2.	Radius of Exposure ....(Figure 7)	ft or m	
3.	Area of Exposure...	ft² or m²	
4.	Value of Area of Exposure		$MM
5.	Damage Factor ....(Figure 8)		
6.	Base Maximum Probable Property Damage – (Base MPPD) [4 x 5]		$MM
7.	Loss Control Credit Factor...........................(See Above)		
8.	Actual Maximum Probable Property Damage – (Actual MPPD) [6 x 7]		$MM
9.	Maximum Probable Days Outage – (MPDO)......(Figure 9)	days	
10.	Business Interruption – (BI)		$MM

(2) For no credit factor enter 1.00. (3) Product of all factors used.
Refer to *Fire & Explosion Index Hazard Classification Guide* for details.

FIGURE 2-9. Dow Fire and Explosion Index *(continued)*. [From *Dow's Fire and Explosion Index Hazard Classification Guide*, 7th ed., 1994. Reproduced with permission of the American Institute of Chemical Engineers.]

(a)   (b)

FIGURE 2-10. Typical Safety Can. (a) View showing spring-loaded cap, which will not open if the can is tipped over. The cap acts as an emergency vent, which opens at about 5 psig when exposed to fire. (b) Cutaway view showing screen, which acts as a flame arrester, mounted on the inside of the spout.

When transporting or using small containers:

- Use proper transport equipment.
  —specially designed drum transport and tilting carts
  —special drum pallets for transport by forklift and for stacking in racks
  —carboy trucks
- Use proper tools to open and close containers.
- Connect grounding cables to containers before attempting to transfer flammable liquids (Figure 2-11). Section 2.2.5.2 discusses static electricity.
- Perform transfers in a restricted area where accidental spills can be safely contained and then cleaned up.

2.4.1.2. Pipeline Transport

When transferring hazardous liquids from tank cars, trucks, or barges, or when handling large quantities of liquid during normal operation, a specially designed piping system should be installed. The most hazardous liquid handling operation is transfer from shipping container to storage. Most spills result from improper procedures or ineffective design or maintenance of transfer equipment.

The liquid transfer system should

- be designed to contain the liquid when properly assembled and operated
- be constructed of materials compatible with the liquid
- withstand maximum operating temperatures and pressures
- contain easily handled leakproof connectors

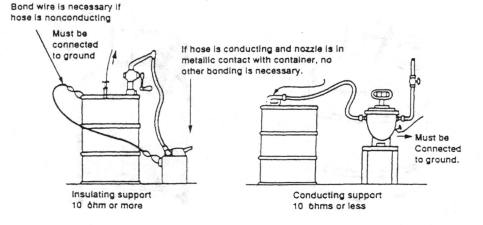

FIGURE 2-11. Bonding and Grounding of Small Containers.

- contain the necessary safety devices
    —relief valves
    —check valves or backflow preventers
    —flame arresters
    —grounding cables
    —heat tracing
    —jacketing
    —vapor return lines
    —ventilated enclosure
- allow transfer to a tank in an isolated area whenever this action will reduce the hazard to the rest of the plant
- contain all necessary vents and drains
- be curbed or diked

Personnel training is also important to prevent accidents. To assure safe operation and the correct responses to emergencies, the following are necessary:

- All involved personnel are familiar with official plant procedures and with safety information provided by the supplier of the hazardous liquid.
- All vents and drains are closed prior to transfer.
- All isolation valves in the transfer line are open; filters and strainers are not plugged.
- All grounding cables are in place on transfer lines subject to static electricity buildup.
- All equipment is operated safely (see Chapter 3).
- A responsible person monitors for leaks during the transfer process.
- Lines are vented, drained, and cleaned as necessary after transfer is complete.
- Correct personal protective equipment (PPE) is available and in proper working order.

With hazardous liquids, dedicated transfer lines should be used as much as possible. There should be a system to identify hazardous piping systems by color coding or tagging with the name of the fluid or the line number. Lettering should be large and should contrast with the color of the pipe covering. It is also useful to place arrows identifying the direction of flow in prominent locations. These line identifiers should be periodically maintained and restored if damaged. In some cases, it will be possible to use unique sizes or types of connections to prevent errors.

2.4.2. Liquid Storage

Hazardous liquids must be stored safely away from process and public areas and in such a manner that leaks or spills can be contained. Good separation criteria can be incorporated into the layout of a processing facility but are not easy to implement for a laboratory or packaging operation that handles a variety of hazardous liquids. The operator must then balance hazardous liquid accessibility against ease of operation and maintenance and should therefore: establish proper conditions for storage

- minimize the quantity of liquid stored
- optimize distance between storage location and point of use
- follow all codes and standards that apply to handling and storage of the liquid (Tables 2-5 and 2-6)
- verify that materials of construction of storage vessels are compatible with the liquid
- establish requirements to maintain contents of the storage vessel in a safe state

—heating
—cooling
—inerting
—venting
  –stand-alone or common header
  –atmospheric or to recovery/destruction
—draining
  –closed system
  –common drain
  –trench
• provide appropriate devices for maintaining safe conditions
  —pressure sensors/alarms
  —temperature sensors/alarms
  —level sensors/alarms
  —dip legs/tubes (see Figure 2-12)
  —manways
  —handholes
  —relief valves
  —rupture disks
  —vacuum breakers
  —conservation vents (see Figure 2-13)

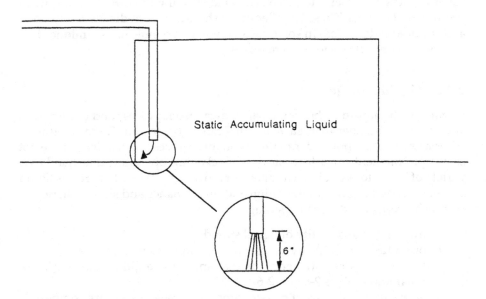

Static Accumulating Liquid

6"

FIGURE 2-12. Proper Installation of a Dip Pipe. A fill pipe entering the top of a tank and carrying a hazardous liquid subject to static accumulation should terminate within six inches of the bottom of the tank and should be installed so as to avoid excessive vibration.

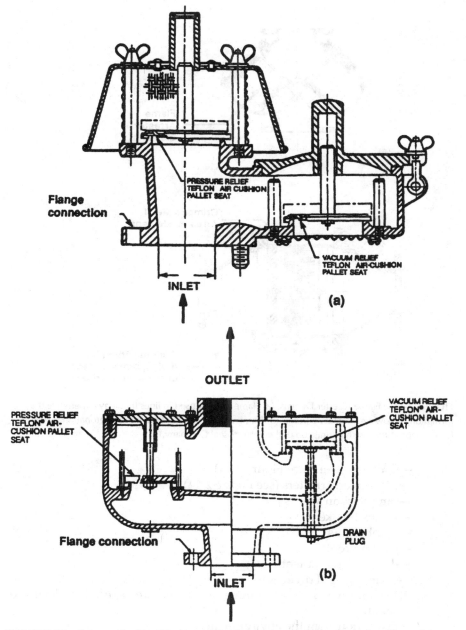

FIGURE 2-13. Conservation Vent Designs. (a) Breather vent mounted on the tank and vented to the atmosphere. (b) Vent mounting for pipe line . (Reproduced with permission of the Protectoseal Company.)

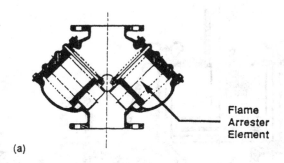

(a)

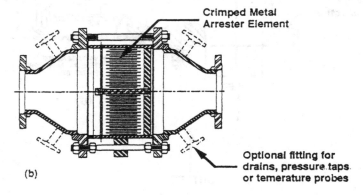

(b)

FIGURE 2-14. Flame Arrester Designs. (a) Flame arrester used in lines that may experience the formation of slow-moving flame fronts. (b) Flame arrester used in lines where detonation waves may form. (Reproduced with permission of the Protectoseal Company.)

> —flame arresters (see Figure 2-14)
> —detonation arresters (see Figure 2-14)
> —sample connections
> —vortex breakers
> —diaphragms
> —floating roofs
> —double containment
> —diking or spill diversion
> • determine whether the storage vessel requires special enclosure or ventilation for
>   —protection from the environment
>   —protection of personnel and the environment from the vessel contents
> • provide instrumentation to warn operators of high liquid level
> • in hazardous situations, provide interlocks to stop filling or divert flow from the tank

2.4.2.1. Large Storage Tanks

Volumes of hazardous materials stored on a site should be kept to a minimum [24], and incompatible materials should be kept separate. The practical minimum storage volume will depend on process demands, the size of the standard shipping container, and the logistics of delivery. Large storage vessels often are needed, and these have their own safety criteria. Standard vessel geometry depends primarily on the volatility of the liquid:

Nominal vapor pressure at ambient temperatures	Storage vessel geometry and design code
Atmospheric to 1 psig	Standard API 650 tanks, vented and fixed roof
	API tanks with inert gas blanketing of vapor space
	API tanks with diaphragm or floating roof
1 to 5 psig	Standard API 620 tanks
5 to 15 psig	Horizontal or vertical tanks with dished heads
	Spherical vessels
	API 620 or ASME Section VIII
15 to 100 psig	Spherical vessels
	Long horizontal cylinders
	ASME Section VIII
greater than 100 psig	Small high-pressure cylinders manifolded together

Most leaks from tanks result from overfilling or vandalism, not from faults in design or construction [25]. Certain tanks will require some type of secondary containment (40 CFR 112). The use of integral, attached steel dikes is becoming more common, and the Steel Tank Institute is developing construction standards [25].

Many tanks require emission controls, and some require emergency pressure relief. These should be in accord with the API code relevant to the tank design and with NFPA 30 and 29 CFR 1910.106. The NFPA also regulates spacing of tanks. Finally, the use of inertion to prevent explosions will make an installation safer and will allow closer grouping of tanks. This is covered in NFPA 69.

2.4.2.2. Small Volume Storage/Area Layout

Small volumes (less than 100 gallons) of hazardous liquid required for laboratory or maintenance facilities should be stored in rooms or small buildings isolated from but close to their point of use (see Figure 2-15). Such liquids should be stored in containers that comply with regulations governing their

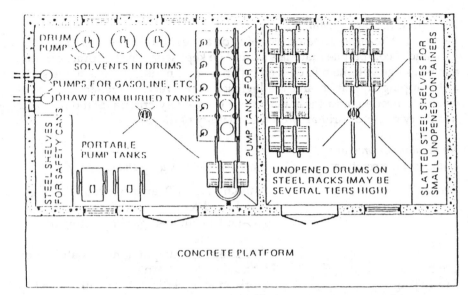

FIGURE 2-15. Storage of Small Containers. Typical layout of a building for storing and dispensing small quantities of flammable solvents. Such buildings should have lighting and electrical systems that comply with the appropriate NEC hazard grouping . [From Fire protection handbook, 16th ed., Figure 11-4]. NFPA, 1986. Reproduced with permission.]

safe handling. For example, volatile liquids should be stored in safety cans such as the one shown in Figure 2-10. Groups of safety cans containing various hazardous liquids should be stored in metal cabinets similar to that shown in Figure 2-16. Tables 2-5 and 2-6 list specific regulations.

2.4.2.3. Underground Storage Tanks
Underground storage tanks (USTs) are not recommended for new installations. Subtitle I of RCRA places restrictions on the use of USTs [26]. New installations intended to hold hazardous materials must have high integrity and conform to Section 4003(g) of RCRA. Where USTs already exist, they require proper safeguards. These may include

- double wall to contain leaks
- containment by surrounding membrane or concrete structure
- monitoring for leaks
- diversion and collection of leakage for proper disposal
- accurate records of tank inventory
- insurance to cover the cost of cleanup after tank failure

Regulations covering USTs that contain hazardous materials are in 40 CFR 280.

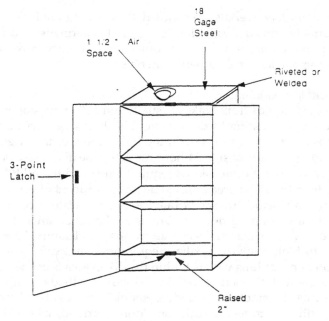

FIGURE 2-16. Metal Storage Cabinet.

2.4.2.4. Inerting

Reducing the oxygen concentration below that which supports combustion will reduce the hazard of flammable liquid storage. Therefore some method of inerting the vapor space with a suitable gas often should be provided. Carbon dioxide and nitrogen are the common choices. Inert gases such as helium and argon are used in special cases but are expensive. Steam is occasionally used for liquids stored above 170°F but presents thermal hazards. Its use can lead to generation of vapor or to the formation of vacuum as the apparatus cools. Inert gases always present the hazard of asphyxiation. Their proper use is discussed in Section 2.6.6.

2.4.3. Spill Control and Cleanup

Spills and leaks can occur during the normal liquid transfer process. Transfer should then stop immediately and resume only after the leak is eliminated and the spill cleaned up. Typical sources of leaks and spills are

- piping flanges
- hose connections
- valve packing
- pump packing and mechanical seals

When hazardous materials are handled, these sources of leaks should be located in curbed or diked areas to localize spills. If a permanent curb or drain can not be provided around the leak source, a temporary curb should be provided, using barriers or absorbent materials.

2.4.3.1. Liquid Impounding

Spillage of stored liquids from tanks is usually controlled by proper drainage or dikes. Vessels containing hazardous liquids should be surrounded by a dike, which is usually sized to contain the full volume of the largest tank it surrounds, along with the design 24-hour rainfall if the diked area is exposed. Figure 2-17 shows some dimensional requirements, and the NFPA [22] also publishes criteria for diked areas. Dikes can be constructed of earth, concrete, or steel. They may include trapped drains, to remove water from rain or fire fighting, and oil separating devices. Drains in the diked area should have shutoff capability to localize spills and allow orderly cleanup. Access should be provided to tanks, valves, auxiliary equipment, and safe exits. Possible accumulation of hazardous vapors within the diked area must be checked.

If containment of the total volume by the above methods is impractical or hazardous in itself, remote impounding should be considered (see Figure 2-18). One method is to locate tanks on sloping ground, where directional dikes or drainage ditches carry the liquid to an impounding basin, away from sensitive areas. The impounding area should be large enough to contain the full volume of the largest tank supported by the drainage system plus accumulated rainfall.

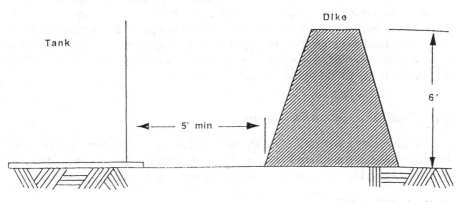

FIGURE 2-17. Standard Dike Arrangement. Height is usually restricted to six feet. When the dike is higher, the distance between a tank and the toe of the interior of the dike wall must be at least five feet.

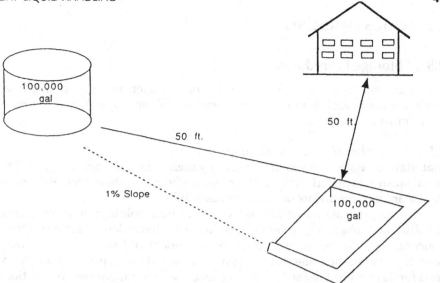

FIGURE 2-18. Use of Remote Impoundment. The impounding area must be large enough to hold the contents of the largest tank that can drain into it. A 15% slope is usually adequate for drainage control.

2.4.3.2. Small Container Leaks

Small containers leaking hazardous liquids should be emptied immediately. The contents should be drained by gravity, siphon, or portable pump into a more suitable container. If immediate transfer is unnecessary or impractical, the container can be placed in another receptacle such as one shown in Figure 2-19. The leaky container can then be taken to an area where liquid transfer and cleanup can be performed safely.

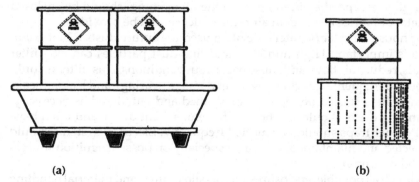

(a)   (b)

FIGURE 2-19. Secondary Containment Units. (a) Secondary containment pallet used to transport leaking drums. (b) Single drum secondary containment.

2.5. SOLIDS HANDLING

2.5.1. Storage Procedures

Solids are usually stored in bulk or piles, or on pallets or racks. In any case, incompatible materials must be separated and all appropriate safety procedures must be followed.

2.5.1.1. Combined Storage and Separation

Materials to be stored are characterized by their degree of hazard or risk. The most hazardous materials must be segregated from the others. One of the most important measurements of risk is combustibility.

Special precautions are necessary with combustible and unstable chemicals that are subject to decomposition or other hazardous reactions. Other concerns include the catalytic effect of containers and materials that could react together in the same storage area. Each stored substance must be examined for its degree of hazard, including severity—type and power of reactions (what can it do?) and sensitivity—initiation modes (what can cause it?)

A hazard assessment also includes normal and failure modes for operating equipment and nonoperating elements as well as environmental interactions. Countermeasures (containment, quenching, dumping, venting, isolation) must be considered for each of the hazardous materials.

2.5.1.2. Bulk Storage and Handling

Dust usually results when solids are conveyed or transferred. High-velocity solids streams produce large movements of air. Suspended dust can be carried long distances, and so dust containment and collection systems are necessary. Collection systems require sufficient air velocity to capture particles and avoid settling and plugging. Air flow can be balanced by air inlet dampers and blast gates.

The filling of a bin with material dropped from a conveyor generates dust. To limit dust escape, the bin (other than the fill opening) should be enclosed by a metal or fabric skirt and an air return duct should be provided.

Bag houses and fabric filters are often used to capture dusts. Their maintenance is important. High humidity and fine dust particles can clog filter materials, reduce air flow, and cause hazardous conditions. Dust filters should be inspected frequently for clogging, filter damage (tearing), and leaking seals. Bags and filters should be replaced or cleaned and laundered as necessary. Surrounding ducts should be checked for plugs, bent ducts, and open blast gates. Powder traps should be emptied frequently. Electrical systems should be inspected and maintained, focusing especially on possible ignition sources and good grounding.

As much as possible, enclosures such as silos, tanks, and material handling structures are designed to relieve overpressures due to explosions. Design provisions include doors, windows, relief panels, and light gauge coverings over the steel structure. All internal surfaces (including structural beams), bottoms, and transfer elements should be self-cleaning in order to avoid accumulation of dust. Vertical walls and overhead structures should be able to be flushed with water periodically. Silos that discharge by gravity can plug by bridging or arching of the product. Silo outlets should be designed so that the material stored does not arch, either mechanically (interlocking of particles) or cohesively (related to bulk strength).

The detection and repair of faulty mechanical equipment and components are important to minimizing ignition sources. Such sources can include hot bearing sensors, speed indicators, alignment devices, level sensing gauges, slowdown detectors, overflow alarms, and pressure gauges. Scrap metal and other foreign materials are removed by screens, grates, scalpers, and magnets.

Flow-promoting devices are often used to allow flow to resume in blocked hoppers. These include air lances (perforated pipes) near the outlets, air injection at hopper sides, air cannons (explosive charge of compressed air) in blocked zones, and mechanical vibration.

When new materials are to be stored in old bins, the strength of the structure should be checked (denser materials will produce higher unit loads). Further, the cone bottom must be checked for sufficient slope to permit free flow and prevent arching. If any of the above flow-promoting devices are available, a blockage might break without manual intervention. If not, an arch can be broken manually from the top or the bottom of the bin. A worker with long tools, working from the top, can often break an arch without having to enter the bin; working from the bottom is more dangerous.

Workers can fall into open storage bins, silos, hoppers, or tanks and thus sustain injuries. To avoid such injuries, bin openings up to two feet above the floor level should have standard guard rails and toeboards around the bin or a cover such as mesh, grating, or parallel bars over the opening. Entering a bin for operating reasons or for repair and maintenance is hazardous. The appropriate safeguards appear in Section 4.5.1 on confined space entry.

2.5.2. Transfer Procedures

Charging of solids into reactors and transfer of solids between operations are among the most important and widely used operations in the chemical industry. The transfer of solid materials often generates dust. This section reviews safety and control implications related primarily to dust generation and flammable atmospheres in such transfer operations.

2.5.2.1. Dust Fires and Explosions
Dust fires and explosions result when particulate materials dispersed in air at certain concentrations are exposed to a source of ignition. An exothermic reaction can proceed to a violent explosion in which large amounts of heat and hot gases are generated rapidly.

When particles fall freely from one level to another, dust clouds may form. The resulting concentration depends on factors such as

- rate of material flow
- height of free fall
- particle size and size distribution
- size of surrounding free volume
- degree of air circulation in and around equipment

These factors are important in relatively passive equipment, such as storage silos, mixers, and settling chambers. More active equipment, such as crushers, grinders, cyclones, dryers, bagging and packaging equipment, and certain conveyors can produce much more dust.

Solid materials that can support an explosion include chemicals, pharmaceuticals, dyestuffs, coal, plastics, wood, foods, agricultural products, and metal powders. Dusts of these materials in air can be explosive, and there are some substances that do not even need air (e.g., some metal powders in nitrogen or carbon dioxide).

To explode, dust must be concentrated within a certain range. Below the lower explosive limit, the heat generated by the combustion of a particle does not reach its neighbors. In theory, above an upper limit, combustion of particles is incomplete, and an explosion is not propagated. However, combustion reactions tend to be surface-area limited, and reported upper limits may not be reliable. The safe approach is to assume that there is no reliable upper limit. Variables affecting the explosive limit include particle size and shape, size distribution, moisture content, electrical conductivity, and humidity of the air. Transfer operations can produce large dust concentration gradients, and so localized ignition is often possible.

The severity of an explosion relates inversely to particle size. Fine powders have large surface areas, stay airborne longer, and require less energy to ignite.

An explosion also requires an ignition source. Such a source can be friction (in a bucket elevator, belt or screw conveyor), welding or cutting, hot surfaces (radiators, steam pipes, electrical appliances), static electricity (solid materials handling equipment such as belt conveyors, pneumatic transport, etc.), direct flame (smoking materials), sparks (friction between metal parts, electrical faults), self-heating (autoreactive chemicals, hot fine powder), or others such as lightning, foreign material, or impact on certain paints. Transported powders and dusts, by their accumulation and discharge or static electricity, can be their own sources of ignition.

Explosion detection and suppression systems can be used for quick response to a developing emergency. They will be effective only if located and installed properly. Emergency vent system design depends on the nature of the dust and the volume and configuration of the system. NFPA 68 covers this subject and gives hazard ratings for certain dusts, along with methods for calculating required vent areas.

2.5.2.2. Health Hazards

Dust inhalation, swallowing, or direct contact with skin, eyes, or mouth can be hazardous. The most common and potentially dangerous is inhalation, whose hazardous nature can depend on particle size. Larger particles are caught or exhaled, but particles less than 0.1 μm can enter the lungs and potentially the bloodstream (see Figures 2-20 and 2-21). They can also be ingested and, if inherently toxic, produce systemic effects.

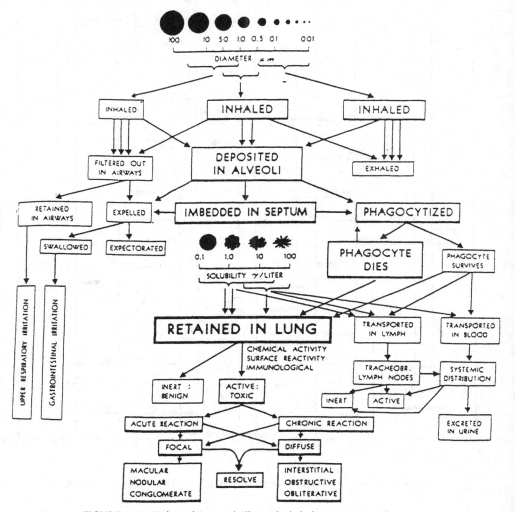

FIGURE 2-20. Biological Fate and Effects of Inhaled Inorganic Particles.

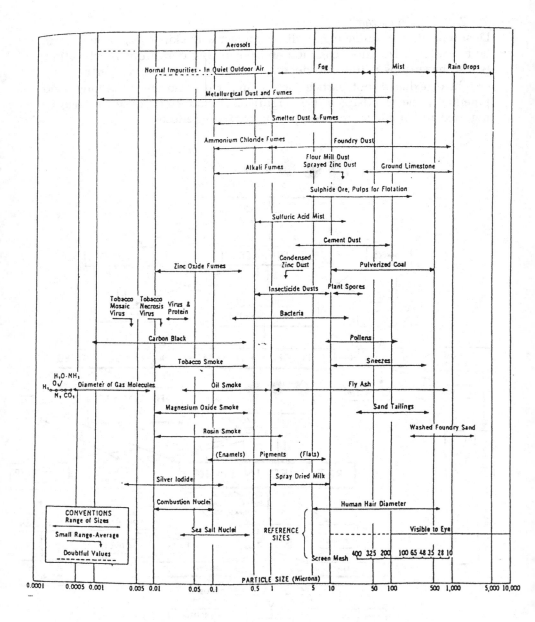

FIGURE 2-21. Sizes of Airborne Contaminants. [Courtesy of Mine Safety Appliances, Co.]

Diseases of the lungs caused by dusts include silicosis (from quartz dust), asbestosis (from asbestos) and cancer. Adverse effects on the blood and blood-forming system can be caused by lead, cadmium oxide, manganese, chromates, arsenic, insecticides, and radioactive dusts.

2.5.2.3. Dust Control

Proper exhaust and venting systems will control dust. Dust collectors, seals, and dust control equipment are used together with designs that minimize free-fall distances, free volumes around dust-producing operations, and accumulation surfaces, while allowing easy cleaning.

Dust control equipment includes collection hoods, fabric and fiber filters, reverse jet filters, wet and dry precipitators, cyclones, and scrubbers [27]. Selection depends on particle size and density of the material, volume and temperature of air to be treated, efficiency, economic considerations, and dust characteristics. Dusts that are sticky, fluffy, erosive, toxic, or combustible demand special consideration. The efficiencies of various types of collectors have been tabulated, along with a guide to selection for various applications [28]. This information is summarized in Tables 2-8 and 2-9. A combination of devices often will be more effective than a single type.

Dust control includes both external and internal methods. Source ventilation, hood ventilation, and room ventilation are typical external methods. Internal methods include reactor ventilation and the use of a closed charging system, which provides a suction transfer of powder into an evacuated reactor. Dust control equipment should be interlocked with production machinery to assure simultaneous operation and protection.

Whenever flammable substances are transferred from one vessel to another, the two should be electrically bonded in order to equalize electrostatic potential and prevent static sparks. Permanent or temporary bonding lines are used to make the metal-to-metal connection. If clamps or clips are used, paint, scale, and corrosion must be penetrated so that bare metal contact is made.

Equipment such as pumps, mixers, process vessels, tanks, and drum racks must also be grounded. Piping should not be used as a conductor unless all joints are welded, soldered or jumpered.

2.5.2.4. Flammable Atmospheres

Risk of fire or explosion increases when solids are charged into flammable atmospheres. Some techniques recommended to avoid these potentially dangerous operations include:

- Maintain an inert atmosphere, whenever possible. If an inert atmosphere cannot be maintained, take precautions such as using acceptable materials of construction for containers and reactors, bonding and grounding, and observing constraints related to flash points of the materials being handled.

TABLE 2-8
Efficiencies of Dust Collectors

Type of Collector	% Efficiency			
	at 10µm	at 5µm	at 2µm	at 1µm
Inertial collector	30	16	7	3
Medium-efficiency cyclone	45	27	14	8
High-efficiency cyclone	87	73	46	27
Low-resistance cellular cyclone	62	42	21	13
Tubular cyclone	98	89	77	40
Irrigate cyclone	97	87	60	42
Self-induced spray deduster	98	93	75	40
Spray tower	97	94	87	55
Wet impingement scrubber	>99	97	92	80
Disintegrator	99	98	95	91
Venturi scrubber-medium energy	>99.9	99.6	99	97
Venturi scrubber-high energy	>99.9	99.9	99.5	98.5
Electrostatic precipitator	>99.5	>99.5	>99.5	>99.5
Irrigated electrostatic precipitator	>99.5	>99.5	>99.5	>99.5
Shaker-type fabric filter	>99.9	99.6	99.6	99
Pulse-jet fabric filter	>99.9	99.6	99.6	99.6

- Charge solids into inerted reactors before adding flammable solvents.
- Provide "closed" charging of solids without breaking the inert atmosphere.
- Use pre-slurry kettles.
- Use dedicated equipment for preparing charges.

Note that expensive control devices do not have to be installed on all reactors (i.e., dust controls), or on mechanized equipment such as drum and bag manipulators.

2.5.3. Bulk Conveying

Solids may be conveyed mechanically or pneumatically. Mechanical conveyors are typically belt, flight, screw, roller, or "en masse" transport devices. Solids are also moved by cranes and bucket elevators and by vibratory, star, and rotary plow feeders.

TABLE 2-9
Guide to Selection of Dust Control Equipment

Factors	Cyclones	Wet washers low energy	Wet washers high energy	Dry electrostatic precipitators	Wet electrostatic precipitators	Aggregate filters	Fabric filters	Fibrous filters
DUST PROPERTIES								
High inlet burden	✓	✓	✓	care	care	care	✓	beware
Erosive	✓	✓	care	✓	✓	✓	care	✓
Sticky	beware	✓	✓	care	✓	beware	beware	care
Light fluffy	beware	✓	✓	care	✓	✓	✓	✓
Difficult to wet	✓	care	care	✓	care	✓	✓	✓
Pyrophoric	care	✓	✓	care	✓	✓	beware	care
Resistivity problem	✓	✓	✓	beware	care	✓	✓	✓
GAS CONDITIONS								
Constant pressure drop	✓	✓	✓	✓	✓	care	care	care
Varying flow	care	care	care	care	care	✓	✓	care
Explosive, combustible[a]	care	✓	✓	beware	beware	care	care	care
Corrosive	care	care	care	care	care	care	care	care
Suitable for high pressure	✓	✓	✓	care	care	✓	care	✓
OTHER FACTORS								
Minimum ancillary equipment	✓	care	care	care	care	care	care	✓
On-line regeneration	✓	✓	✓	✓	care	care	care	beware

KEY
✓ Indicates that the type of plant can generally cope with the process requirement, if well designed.
care Indicates that special attention is required in plant design and operation to prevent problems.
beware Indicates that this process condition could lead to severe operational difficulties.
Alternatives that avoid the problem are normally sought.

[a] For all gas cleaning problems associated with explosive or combustible materials, competent advice should be sought.

The choice of transport equipment is based on several considerations:

- rate of material flow
- distance and inclination of travel
- environment through which material travels
- lump size, density, flowability, abrasiveness, toxicity, corrosiveness
- degree of potential dust generation

For example, screw and "en masse" conveyors can handle only small lumps. Belt conveyors are useful for long distances but not for high temperatures. Vertical lifts often require bucket elevators.

Hazards of concern are dust explosions, the escape of flammable or toxic dusts, and injury from contact with moving machinery. Loading and unloading sections are areas of particular danger. Bucket conveyors often have materials jammed between the buckets and the side of the boot. Chutes may clog because of accumulation of material on their side walls.

Loading and discharge points of conveyors must generally be ventilated and covered with hoods, since it is usually not possible to provide dust-tight enclosures or use wetting to reduce dust.

Temperature is a major consideration from a fire protection standpoint. Flame propagation must be avoided. Belt conveyors are usually restricted to 150°F. Rubber belts have the highest rate of flame propagation, while the ignition characteristics are similar for neoprene, rubber, and polyvinyl chloride materials. With hot materials or in hot atmospheres, screw conveyors, vibrating units, or types of pan conveyors are preferred.

2.5.3.1. Mechanical Conveyors

Automatic sprinkler or water spray protection is generally needed for major belt conveyors, especially if they are enclosed. Controls should be provided to shut down the conveyor whenever the sprinklers are activated. Hose streams should be able to reach all portions of the conveyor.

To prevent dust clouds, choke feeds should be provided when loading dusty material onto conveyors, and sharp changes in direction of conveyors should be avoided. Dusty materials must be ventilated at changes in direction and at the ends of lines to remove and collect dust. Conveyor piping should be strong enough to resist the pressure of a dust explosion.

Conveyors such as screw, drag, and "en masse" types should be fully and tightly enclosed, with access covers at the discharge end and for shaft couplings. Access covers must be securely fastened to prevent the escape of dust.

Mechanical conveyors running between buildings or through firewalls should have suppression systems or special chokes or seals to prevent the propagation of an explosion.

Dust collectors should be located away from process equipment, either outside or in separate rooms with adequate explosion venting. Bag collectors should be in metal housings. Where practicable, each process should have its own collector to avoid hazardous mixing.

Conveyors and machinery, including pulleys, guards, and other metallic components, must be bonded and grounded to prevent static discharges. Belts should be made of a conductive material, have a conductive coating, or have a grounded static collector just beyond the point at which the belt leaves the pulley.

The transport of hot or dry materials, or transport through hot or dry air, can increase the generation of static charges.

Water spray or fire doors can be actuated by heat to protect conveyor openings in walls and floors. Water spray, using nozzles directed at the opening, provides pressure and cooling to hot gases. Fire doors can be used to close off the opening, but the actuator system must assure that obstructions will be avoided or cleared automatically.

A fire hazard is created by the accumulation of grease and dirt and the overheating of defective or unlubricated parts such as rollers. Proper inspection and maintenance are necessary.

2.5.3.2. Bucket Elevators
Bucket elevators must be properly designed, installed, operated, and maintained to avoid hazards of high temperature, friction, and overheating. Dust control is necessary with many materials, and collection and venting are important. Doors and casings must be dust-tight, access for cleaning and lubrication must be adequate, and explosion venting must be provided when demanded by the material being handled. At the same time, protection against improper opening of an operating system is required. This can include interlocking closures with the drive and providing emergency stops at each level though which the elevator travels. Jamming or clogging should be sensed automatically and the system shut down to avoid damage.

2.5.3.3. Pneumatic Conveyors
Pneumatic conveyors may operate under pressure or suction, even within the same system (e.g., push/pull conveying). Both types are subject to similar hazards. Flammable materials are of particular concern and should therefore be transported by an inert gas. Inert gases should contain no carbon monoxide and have dew points low enough so that free moisture cannot condense or accumulate in the system.

Some limitations on the transporting gas are material-dependent. For example, the inert gas for magnesium dust should not contain carbon dioxide. Aluminum powder has been transported using controlled flue gas with 3 to 5% oxygen. Inert gas systems should operate under pressure to prevent inleakage of air. Oxygen monitoring is recommended, so that action can be taken if levels become unsafe. When hazardous material is conveyed, the efficient recovery of fine particles from the air discharge is particularly important.

To avoid dust explosions, pneumatic conveying systems should be checked for:

- Air tightness: If pressurized, to prevent the escape of dust from the system; if under negative pressure, to prevent air or contaminants from entering the system.
- Physical strength: To remain intact and tight under normal operating conditions, including vibration; also, under certain circumstances, to withstand or contain explosive pressures.
- Electrical conductivity and grounding: To drain off static electric charges due to material flow; can include bonding across joints, electrically conductive bags for bag filters, grounding for electrically isolated metal objects within the system, and wire braid within rubber-covered transfer hoses.
- Safe electrical installation: To meet electrical classification imposed by conveyed materials and the surroundings.
- Safe construction materials: To be compatible with conveyed materials and the surroundings; also use of special materials, such as nonferrous metals, to minimize spark generation if moving parts should contact one another.
- Filtering systems: To eliminate foreign materials by means of screens, magnets, or metal detectors and extractors.
- Maintenance and inspection program: To check alignment of drives, clearances, dust tightness, electrical bonding and grounding, and control of ignition sources. Suitable joints and access openings must be provided for maintenance, cleaning, and unplugging.

Explosion relief vents for conveyor ducts should have sufficient areas or openings and be protected by antiflashback swing valves. Fans should be constructed of conductive, nonsparking materials and should be downstream of dust collectors. Personnel should not be within fifty feet of the fan during operation. The system must be periodically shut down for inspection and maintenance, including cleaning of accumulated dust. If the fan must be approached during operation, such as for a pressure test, competent technical personnel must direct the efforts and operating management must approve the work.

Separators for air and material should be located outside the building, have lightning protection, and be electrically conductive and bonded. They should be made of noncombustible materials and equipped with filters of low-hazard or flame-retardant substances in metal enclosures. Material discharge outlets should have positive choke devices. Exhaust air should be directed outdoors or recirculated into the pneumatic system. Sight glasses are not recommended as routine. If necessary, they must be noncombustible, rated for the appropriate pressure, sealed, and installed without internal surface discontinuities.

2.5.3.4. Conveyor Standards

There are standards for the various types of conveyor. These are published by ANSI and the Conveyor Equipment Manufacturers Association (CEMA) as CEMA standards. These should be consulted by the purchaser and user of conveying equipment.

Besides the common types of conveyor, CEMA standards also address classifications and definitions of bulk materials. CEMA 550, for example, presents an alphanumeric method for classifying bulk density, particle size, flowability, abrasiveness, and handling characteristics of solids. The identifier contains, in order:

- a two digit number which gives the loose bulk density in pounds per cubic foot.
- a letter and subscript numeral describing the particle size
- a single digit describing the flow characteristics
- a single digit describing abrasiveness
- a series of letters identifying hazards and properties that can make handling of the material difficult

This system is explained in the standard, and the conventions for each element of the identifier are listed.

For example, sodium sulfite is described by $96B_646X$ and pebble lime by $55C_{1/2}25HU$. These show that sodium sulfite has a normal bulk density of about 96 lb/ft^3; lime is coarse, with a particle size of about ½"; sodium sulfite flows sluggishly (4) while pebble lime is relatively free-flowing (2); sodium sulfite is moderately abrasive (6) while lime is only mildly so (5); sodium sulfite packs under pressure (X) and may become more difficult to handle; and lime is hygroscopic (U) and may decompose or deteriorate in storage (H).

Other standards cover standard terms and definitions (CEMA 102) and dimensional standards (e.g., CEMA 300 for screw conveyors). Other publications describe installation standards and conveyor performance guidelines.

2.5.4. Solids Packaging

2.5.4.1. Packaging Materials Selection

Industrial packaging is concerned primarily with the safe and economical transportation of goods to their destinations, and with their protection during handling and warehousing.

Proper selection of packaging materials depends on:

- Substance being packaged: physical/chemical properties, quantity, degree of hazard, stability, etc. Packaging of hazardous materials will require special considerations.

- Common-carrier regulations, whether shipment is over the road or by railway, ship, or air. DOT regulations contain specific requirements for shipping packages.
- Durability of container, adequacy of interior packaging, and protection given to contents by packaging (cushioning, clearances, support, positioning). The most nearly noncombustible suitable containers should be used.
- Availability of material or combinations of materials needed.
- Specific protective qualities such as resistance to moisture, water, puncturing, scuffing, or abrasion; protection against shock, vibration, insect infestation, odor, grease, acid, light, ambient changes, mold, tarnish, pilferage, and other hazards.
- Strength of component materials and unit as a whole, involving parameters such as burst strength, flat crush resistance, compressive strength, and puncture resistance.
- Other issues that may be important to recipients, including style, economics, labeling, dimensions, tolerances, and means of fastening or sealing.

2.5.4.2. Packaging Operations Hazards

The hazards in a packaging operation reflect the dangers due to the chemicals or materials being packaged, the container or packaging materials, other packing or dunnage materials, and the machinery used (e.g., ignition sources, moving parts, etc.). Hazards particularly related to solids include dust generation and the ventilation required for control.

Most hazards related to packaging and packing materials result from combustibility. Plastic pellets, rigid foams, excelsior, shredded paper, sawdust, burlap, and other such materials are combustible. Large quantities may have to be kept in special vaults or storerooms. Areas where considerable quantities of packing materials are stored or handled should be protected by automatic sprinklers that meet all fire codes. Large amounts of sacking must be protected against spontaneous ignition. Piles should not be too high and should be ventilated, with vents allowing air to flow from the outside into the center. Fire-resistant rooms with sprinklers and dust-tight lights are advisable.

Used or waste packing materials, including crating, should be removed and recycled or disposed of as soon as possible. Packaging materials should not be strewn about the facility.

Otherwise, normal care must be taken regarding:

- disposal of smoking materials
- housekeeping
- packing and unpacking of related goods
- storage of idle pallets
- clearance from heat-producing equipment
- special equipment hazards

2.5.4.3. Labeling Requirements

Packages should be marked with the name of the product and hazard information. NFPA codes and CAER, HAZCOM, and DOT regulations have specific requirements [3, 4, 5, 6]. Special concern should be given to materials that are oxidizing, combustible, unstable, reactive with water or air, corrosive, self-heating, or otherwise hazardous.

Other data normally included are name and address of supplier, net weight, and process information such as batch, blend or lot number, date of manufacture, and other "variable" information. Shelf life information, where applicable, is also appropriate.

The hazard identification system used by NFPA and the warning labels required by DOT for shipping hazardous materials are discussed in Section 2.1. Other warning signs may be used, such as those for machine hazards, protective equipment requirements, emergency procedures, chemical disposal and recovery tags, and other OSHA safety signs and tags.

2.6 GAS HANDLING

This section covers the handling and use of compressed and liquefied gases. These gases are usually characterized by high pressure or low temperature and are therefore supplied in special containers. Compressed gases may be used for atmospheric control, fire protection, fuel, medical application, etc. Compressed and liquefied gases may be reactants, oxidants, refrigerants, or pesticides.

2.6.1. Classification of Gases

1. *Nonliquefied Gases*—do not liquefy at ambient temperatures at pressures up to about 2500 psig. Examples are helium, hydrogen, nitrogen, and oxygen.
2. *Liquefied Gases*—liquids at ambient temperature under pressures of about 25 to 2500 psig. Examples are ammonia, carbon dioxide, chlorine, nitrous oxide, propane, and sulfur dioxide.
3. *Dissolved Gases*—gases carried as a solution in another material. The common example is acetylene, which is dissolved in acetone. The acetone in turn is adsorbed on a porous mass that fills the interior of a cylinder.
4. *Atmospheric Gases*—gases naturally found in the atmosphere, including argon, carbon dioxide, hydrogen, helium, krypton, neon, nitrogen, oxygen, radon, and xenon.
5. *Fuel Gases*—gases intended for burning in air or oxygen. Examples are acetylene, butane, hydrogen, liquefied petroleum gas (LPG), methyl acetylene–allene mixture (MAPP gas), propane, and other hydrocarbons.

6. *Fumigant/Sterilant Gases*—gases used to kill vermin or to sterilize medical supplies and equipment. Examples are carbon dioxide, ethylene oxide, methyl bromide, and propylene oxide.
7. *Refrigerant Gases*—gases that easily liquefy under pressure. Examples are ammonia, propane, certain halogenated hydrocarbons, and sulfur dioxide.
8. *Medical Gases*—gases or mixtures of gases used in medical applications. Examples are air, carbon dioxide, helium, nitrogen, nitrous oxide, and oxygen.
9. *Specialty Gases*—gases used in specialized applications and usually of high purity. Examples are arsine, carbon monoxide, diborane, ethylene oxide, fluorine, hydrogen chloride, mercaptans, phosgene, phosphine, propellant gases, radioactive gases, silane, sulfur dioxide, and vinyl chloride.

2.6.2. Regulations and Standards

2.6.2.1. Industry Standards
The Compressed Gas Association (CGA) is the primary technical association that publishes standards for the safe handling, transportation, and use of compressed gases. CGA also publishes a handbook [29] that is a comprehensive overview and an indispensable reference on handling compressed gases. It is the basis for much of the information that follows. The CGA also publishes technical pamphlets and standards, many of which are incorporated into governmental regulations. Some of these are referred to below by their identification numbers. NFPA standards also apply in certain cases.

2.6.2.2. Transportation
Transportation of compressed gases by air, highway, rail, or water is regulated by the Department of Transportation (DOT). DOT regulations cover classification of containers, design specifications for each type of container, qualifications for containers, filling requirements, maintenance, material compatibility, documentation, and rules for labeling, marking, and placarding. The DOT regulations are published in 49CFR100 through 199 and 390 through 397. Title 49 regulations are republished by the Bureau of Explosives through The Association of American Railroads as Tariff BOE 6000.

2.6.2.3. Safety
The Occupational Safety and Health Administration (OSHA) regulates employee safety and health aspects. OSHA regulations directly concerning compressed gases are in 29CFR1910. Compressed gases can be asphyxiating, corrosive, flammable, reactive, or toxic. Employees who handle, store, transport, or use these gases should be trained and be familiar with all related safety

and health aspects. Material Safety Data Sheets (MSDS) for compressed gases must be provided by the suppliers, who also can often provide useful information on the handling of their products and containers.

2.6.2.4. Fuel Gases
Facilities and procedures for handling fuel gases should follow the National Fuel Gas Code, which is presented in NFPA 54. Specific precautions will also apply to the individual gases. The storage and handling of LPG, for example, are covered by NFPA 58.

2.6.2.5. Fumigant/Sterilant Gases
These gases are regulated by the Environmental Protection Agency (EPA) under the Federal Insecticide, Fungicide, and Rodenticide Act (FIFRA). FIFRA regulations are in 40CFR164.

2.6.2.6. Medical Gases and Devices
The Food and Drug Administration (FDA) regulates medical gases as drugs and certain compressed gases and gas equipment as medical devices. These regulations are in 21CFR201, 207, 210, 211, 801, 807, and 820.

2.6.3. Gas Containers

The subsections that follow treat various types of gas containers. The following precautions apply to all types:

- Identify each container by its label before using. Never rely on color coding, and return to its source any package without a proper label.
- Never use a container for other than its intended purpose.
- Never repair or alter containers or relief devices in an unauthorized facility.
- Never change, modify, or obstruct the discharge ports of relief devices.
- Never use a pressure regulator in other than its intended service.
- Never mix gases in containers or transfer gases between containers.
- Ensure proper labeling of containers and identification of contents; never deface, alter, or remove an identifying label.
- Follow applicable codes in handling containers and in the design, installation, and operation of piping systems.
- Verify that correct pressure regulators are used; be certain that threads fit and never force them.
- Prevent backflow into cylinders; keep systems simple or add positive protection where necessary.
- Open container valves slowly and apply pressure to piping systems gradually.

2.6.3.1. Cylinders

Compressed gases are frequently shipped in high-pressure cylinders with volumetric capacity usually equivalent to no more than 1000 lb of water. Certain classes (described below) may be larger, equivalent to as much as 5000 lb. Cylinders are manufactured to the ASME code and DOT regulations. They require independent inspection before being put into service and then periodic retesting. The most commonly used cylinder types are described in Table 2-10. Each cylinder must bear certain marks set out in 49CFR178. These include

- specification used in manufacture (see Table 2-10)
- design pressure
- manufacturer's serial number
- manufacturer's symbol
- inspector's symbol
- month and year of first inspection
- method used for bottom closure (spun or plugged)

The prescribed retest interval for approved cylinders is usually five years. Some types (3A and 3AA) can qualify for a ten-year interval. Such approval carries extra requirements for maintenance and use of the cylinders. These requirements are covered in 49CFR173.34.

Within certain specifications, some cylinders can be charged up to 10% above the indicated design service pressure. Such overcharging is indicated by a plus (+) sign immediately after the test date. Requirements are in 49CFR173.302.

2.6.3.2. Cryogenic Containers

Cryogenic containers are of special construction, intended to hold liquids at very low temperatures from about –130°F down to absolute zero. These are double-walled vessels, with the annulus evacuated and filled with insulation. Venting of small amounts of vapor provides some autorefrigeration to dissipate heat.

DOT regulations apply when the contents are flammable or when the pressure is more than 25.3 psig.

2.6.3.3. Tube Trailers

High-pressure, nonliquefied gases are often shipped over the road in tube trailers. These trailers have a number of separate long cylinders manifolded into a common header. Newer tubes are built to Specifications 3AX, 3AAX, and 3T.

2.6.3.4. Portable Tanks

DOT defines a portable tank as any package over 110 gallon capacity and designed primarily to be loaded into or on, or temporarily attached to, a transport vehicle or ship. DOT Specification 51 covers welded or seamless steel

TABLE 2-10
Common DOT-Regulated Cylinders

Cylinder Type	Construction	Material	Pressure Rating, psig	Max. Water Capacity, lb	Maximum Dimensions	Notes
DOT 3A	Seamless	Carbon steel	1800–10,000	—	—	
3AA	Seamless	Alloy	1800–10,000	—	—	
3AL	Seamless	Aluminum	150+	1000	—	
3E	Seamless	Carbon steel	1800	—	2" dia × 2'	
3HT	Seamless	Alloy	1800+	150	—	1
3AAX		Alloy	—	—	22" dia × 34'	2
3T	Seamless	High-strength alloy	1800+	—	—	
4B, 4BA,, 4BW	Welded	Carbon steel	240	1000	—	
8, 8AL	Welded	Carbon steel	250	—	—	3
39	Welded, seamless, or brazed	Carbon steel or aluminum	500	55	—	4

NOTES
1. Designed for service inside aircraft.
2. Trailer tubes similar to 3AA; minimum capacity 1000 lb water.
3. For acetylene service only. Must contain permanent monolithic porous filler capable of absorbing specified amount of acetylene solvent.
4. Not refillable. Also covers cylinder with maximum capacity of 10 lb water at pressures above 500 psig.

tanks with more than 100 lb water capacity and service pressures between 100 and 500 psig.

Other specifications, including requirements for qualification, maintenance, and use of portable tanks, are covered in 49CFR173.32. Shipping requirements for compressed gases are in 49CFR173.315, and documentation and marking requirements are in 49CFR172.

2.6.4. Cylinder Auxiliaries

2.6.4.1. Pressure Relief Devices

Pressure relief devices are often installed to prevent catastrophic failure of a cylinder. They are subject to a standard fire test defined in 49CFR173.34d. The Compressed Gas Association's standard CGA S-11 defines a number of de-

TABLE 2-11
Gas Cylinder Relief Devices

Type	Description	Action	Complete Release	Mounting
CG-1	Rupture Disk	Responds to excessive pressure in cylinder	Yes	On cylinder or part of valve
CG-2	Fusible Plug (nominal melting temp. = 165°F)	Responds to high temperature inside or outside of cylinder	Yes	On cylinder or part of valve
CG-3	Fusible Plug (nominal melting temp. = 212°F)	Responds to high temperature inside or outside of cylinder	Yes	On cylinder or part of valve
CG-4	Rupture Disk plus CG-2 Plug	Needs both high temperature and high pressure to respond	Yes	On cylinder or part of valve
CG-5	Rupture Disk plus CG-3 Plug	Needs both high temperature and high pressure to respond	Yes	On cylinder or part of valve
CG-7	Relief Valve	Spring-loaded valve opens when pressure is excessive	No	Threaded onto cylinder

vices, which are considered in Table 2-11. The two major types are rupture devices and relief valves. The former allow complete release of a cylinder's contents when they open. Other limitations are discussed below.

Type CG-1
- High-temperature weakening of cylinder can reduce its strength below the pressure set by the disk.
- External corrosion can have similar effects on the cylinder or can weaken the disk itself, allowing premature failure.

Types CG-2, 3
- Possibility of cold flow of fusible plug (restrict to cylinders rated for 500 psig or less).
- No protection against overcharging of cylinder.

Types CG-4, 5
- No action without combined high temperature and high pressure.

Type CG-7
- High-temperature weakening of cylinder.
- More problems with freezing, sticking, etc., of relief device.

In all cases, the material of construction of the safety device must be compatible both with the fluid in the cylinder and with the valve or cylinder wall in which it is mounted. Several rules apply to installation and maintenance of relief devices:

- Use only original equipment manufacturer's parts for repair or replacement (exception: parts whose inter-changeability has been demonstrated by suitable test).
- Allow only trained personnel to install or service relief devices.
- Follow manufacturer's instructions precisely during assembly and installation (e.g., using a torque less than recommended may produce an inadequate seal, while too high a torque can stress the relief device and allow release at a pressure lower than desired).

2.6.4.2. Cylinder Valves

The CGA has defined standard valve connections for most of the gases that are commonly shipped in cylinders. CGA V-1 defines these connections, and the recommended standards are identified and tabulated in Chapter 8 of the CGA Handbook. Along with identification of the gas, the Handbook indicates the volume and pressure of the cylinder, the type of connection, and whether the connection is for withdrawal of liquid. The "Standard" connections are always preferred; these are the CGA's recommendations for the specific gases. "Limited Standard" connections may be recommended with some limitations on their use. "Alternate Standards" were to be phased out by the beginning of 1992. CGA recommends that cylinders fitted with these valves are not acceptable, and the user should change the valves to meet the standard [30].

When a cylinder contains a mixture of gases, valve selection criteria are more complex. Each component of the mixture has its own hazardous properties, and two different components can present two different types of hazard. The standard procedure for choosing valve outlet connections in such cases is covered in CGA V-7, which considers the fire hazard, toxicity, state, and corrosiveness of each component. Most of the mixtures of interest to readers of this volume are subject to CGA V-7; an exception is mixtures of medical gases, which are covered instead by CGA V-1.

2.6.4.3. Pressure Regulators

For safe withdrawal and use of a gas, its pressure usually must be reduced below that in the cylinder. The choice of a regulator to accomplish such reduction is influenced by several factors:

- composition of the gas
- cylinder pressure
- range of delivery pressures
- required flow rate
- degree of accuracy to be maintained in delivery pressure

The active element in a regulator usually is a diaphragm loaded by a spring or with a gas. The diaphragm will open when the downstream pressure falls below a preset value. The body of a regulator usually carries two gauges; one shows the pressure at the source (usually a cylinder), while the second shows the downstream set pressure. The latter can be varied by adjusting tension on the spring or changing the pressure of the loading gas.

For simple applications, single-stage pressure regulators will suffice. Where it is necessary to maintain delivery pressure over wider ranges of cylinder pressure and gas withdrawal rate, two-stage regulators may be required. In any case, the regulator must be chosen for the specific gas or mixture and must fit the cylinder valve properly. These valves are regulated by national standards and are covered in CGA V-1.

The following principles apply to automatic pressure regulators used on gas cylinders:

- Never force the threads when attaching a pressure regulator to a cylinder. Lack of easy fit may indicate that the wrong regulator is being used.
- After mounting the regulator, rotate the delivery pressure adjusting screw counterclockwise (towards zero pressure) until it rotates freely.
- Open the cylinder valve slowly and carefully, and check that the pressure is close to the expected value.
- Be sure that the flow-control or isolating valve is closed, and then turn the delivery pressure adjusting screw clockwise until the desired pressure is reached.
- Never use a regulator as a flow-control device.

2.6.5. Cylinder Handling Procedures

2.6.5.1. Before Filling

A cylinder should not be filled until certain inspections are made and DOT and ICC requirements met (49CFR173.34). Depending on the nature of the gas, there may be other industry standards or recognized good practices to observe. All relevant requirements should be assembled in a set of written procedures that are audited and enforced.

DOT requires first that ownership of the cylinder be determined. If it is not the property of the company filling it, there must be written authorization from the registered owner. The cylinder markings described in Section 2.6.3.1 should be examined and verified as suitable for the lading. These markings will indicate the cylinder specification, the approved service pressure, and the latest test date.

A cylinder must not be filled until the date of its last hydrostatic test has been checked. If the time allowed between tests has expired, the cylinder must not be filled until the prescribed tests are made. Test intervals, methods of hydrostatic testing, and requirements for approval of testing equipment are covered in 49CFR173.34, CGA C-1, and CGA P-15. DOT requires external and internal visual examinations along with each retest. Such examinations are best done before the hydrostatic test itself and should cover:

- external damage
  —burns
  —exposure to excessive heat
  —pits
  —distortion
  —mechanical damage
- valves
  —suitability for lading
  —free of hazardous contamination (e.g., oil on oxygen cylinders)
  —threads at cylinder connection exposed to allow examination
  —handwheels and stems undamaged and in working order
  —smooth operation
  —outlets and threads not damaged
- pressure relief devices
  —in accord with CGA S-1.1
  —no signs of abuse, damage, or extrusion of fusible metal plugs
- neckrings
  —threads not worn or damaged
  —no distortion or movement from position
- cleanliness
  —to suit lading
  —in accord with DOT regulations, company standards, and industry group recommendations.

Many of the observations included above are subjective or at least judgmental. Experience becomes a very important factor. An organization lacking this expertise may wish to contract with a certified inspection firm.

The results of the inspection, including the recommended disposition of the cylinder, should be recorded. Typical contents of a report form might be:

- Cylinder identification
  —Manufacturer
  —Date of manufacture
  —Serial number
  —Symbol
  —ICC/DOT specification
- Protective coating
  —Type
  —Condition

- Inspections (checklist)
  —Corrosion
  —Dents
  —Cuts and gouges
  —Bulges
  —Leaks
  —Fire damage
  —Neck
  —Attachments
- Disposition
  —Date of inspection
  —Name (initials) of inspector
  —Disposition
    –Return to service
    –Repair
    –Scrap

The topic of inspection and hydrostatic testing is covered in detail in Chapter 9 of the CGA Handbook.

2.6.5.2. Cylinder Filling
Safe filling of cylinders demands a number of measures beyond assuring that the cylinders themselves are properly chosen and constructed. Essential steps in the filling of any cylinder include:

- connection to fill line or manifold
- inspection and testing of piping, pigtails, flexible hoses, etc.
- filling procedure (CGA P-15 Appendix 1)
- inspection during or after pressurization
- shutoff and disconnection
- venting of cylinder and connections
- leak testing (connections, valves, relief devices)

Instructions for filling should include:

- purging
- checking
- venting
- disposal of vent
- charging—weighing, pressure measurement
- checking—analyze components; certify; quarantine
- labeling/placarding—DOT; industry standards; certificate; color code
- handling/return instructions
- QC on return

Liquefied Gases

Cylinder-filling techniques for liquefied and nonliquefied gases are quite different. With liquefied gases, cylinder filling is by weight. Scales should be calibrated at least yearly and checked daily. Supports and piping connections must not impede free vertical movement of the cylinder as it is filled.

Each cylinder should be stamped with its tare weight (without valve cap). This weight should be checked by mounting the empty cylinder on a scale. Excess weight usually shows that the cylinder is not empty. A low weight may be due to error, damage, or corrosion. If the user has no previous record on the cylinder, it should be returned to its source.

Cylinders may be evacuated to help the filling process. In the case of a liquefied gas, this introduces the potential problem of fractionation. High boilers can accumulate as the major component vaporizes. These materials may be:

- impurities, which eventually can build up above safe or specified levels
- additives (e.g., inhibitors), which may be unstable at high concentrations or become impurities of concern
- unstable compounds, which can reach dangerous concentrations (peroxides are the most familiar examples)

Such fractionation is a highly specific situation, and corrective action will depend on the nature of the contents.

Only one cylinder should be filled at a time. While the gross weight is monitored, the cylinder is filled to the appropriate loading. Using the known volume or the water capacity of the cylinder, the maximum acceptable net weight is set by the prescribed filling density for the gas (49CFR173.304). Once the cylinder is filled and disconnected, the gross weight should be checked promptly. An overfilled cylinder should be vented immediately to a safe weight.

Finally, the fill line should be emptied of liquid, in order to prevent overpressure due to expansion. The cylinder itself should be checked for leaks (valves and relief device).

Nonliquefied Gases

Because no change in phase occurs, it is easier to load nonliquefied gases. Written procedures should cover the seven major steps listed at the beginning of this subsection. Checklists and periodic reviews are useful.

The procedures must also cover the source of the gas. If a compressor is used to boost the gas to filling pressure, its startup, operation, and shutdown must be covered. Backflow prevention must be designed into the system, along with protection against overpressure at the source or the cylinder.

Use of manifolds and simultaneous loading of a number of cylinders are common practices with nonliquefied gases. Temperature will usually rise as a cylinder is filled, because of the heat of compression of the gas. This rise is detectable on the cylinder walls by hand. When several cylinders are filled at the same time, any not warm to the touch should be valved off and removed from the filling system for diagnosis.

The gradual increase in cylinder pressure allows intermediate pressure testing of connections. 300 psig is a frequently used level.

2.6.5.3. Transfer Operations

Transfer of gases into or out of a process or from one apparatus to another can pose special hazards. All the issues concerning compatibility of the fluid with its container apply to the new combination. The heat of compression of the gas as it enters the new system is sometimes a special consideration. Other potential hazards include:

- leaks
- unintended reaction with the new system or its contents
- lack of intended reaction, leading to excessive accumulation of gas

A gas is frequently added to a system in which it should react or be absorbed. Such a process usually depends on a mass-transfer device or an outside source of energy. If the device is faulty or the energy input missing, the process will fail. Gas can accumulate to high partial pressure or can pass through to downstream equipment or the atmosphere. Typical process malfunctions are:

- agitator failure
  —mechanical damage
  —electrical outage
  —slippage in transmission
- sparger malfunction
  —physical break
  —loose connection
- inadequate supply of absorbent
  —low volumetric flow
  –pump failure (mechanical or electrical)
  –line restriction
  –valving error
  —small batch size
  —low concentration of reagent

Some of the above are internal problems that can not always be detected beforehand. The best prevention is a conscientious program of periodic inspection or proof testing and preventive maintenance.

Other problems can be controlled or prevented by analysis and process monitoring. The typical system in which either continuous or occasional releases are handled by a circulating batch of absorbent has been analyzed for the case of chlorine gas [31]. These systems can be highly dependent on operator awareness.

Although there is often a tendency to ignore a system that is not under load, proper monitoring is essential in order to avoid breakthroughs.

2.6.5.4. Labeling and Placarding
All cylinders must be marked in a way that allows the user readily to identify the contents and associated hazards. The markings must be legible, durable, and conspicuous.

Table 2-12 summarizes cylinder loading regulations; the left-hand column identifies basic requirements; other columns list applicable standards and regulations.

2.6.5.5. Handling Guidelines
A list of guidelines, derived from various sources, to the safe handling and use of gas cylinders follows. A two-part series [32] expands on these rules and discusses certain special conditions. Most of the precautions listed below are paraphrased from rules established by the Compressed Gas Association. Some also appear in DOT regulations.

General Precautions
1. Do not drop cylinders, roll them, or permit them to strike any object violently.

TABLE 2-12
Summary of Cylinder Labeling Regulations

Requirement	Regulating Agency	CFR Reference	Other Reference
Proper shipping name	DOT	49-172	—
Hazardous material product identification number	DOT	49-100 to 199	—
Color-coded diamond label	DOT CGA	49-172 Subpart E —	— C-7, Appendix A
Hazard warning andprecautionary information	OSHA (Hazcom) CGA ANSI	29-1910.120 — —	— C-7 Z129.1, Appendix B
Foods & drugs	FDA CGA	21-201, 801 —	— C-7, Apps B & C
Others	DOA DOT EPA OSHA States	(Consult agency)	(Consult agency)

2. Keep valve caps on cylinders at all times except when in use or being charged.
3. Do not use cylinders as supports or for any purpose other than the transportation and supply of gas.
4. Open cylinder valves slowly. Do not use a wrench to open or close a handwheel valve. If a valve cannot be operated by hand, it should be repaired.
5. Before attaching cylinders to a connection, be sure that the threads mate and are of a type intended for the gas service as specified in CGA Standard V-1.
6. Do not convert a cylinder from one gas service to another without first removing the original contents. If residues are present, clean and purge the cylinder appropriately (refer to CGA C-10).
7. Do not permit oil or grease to come in contact with cylinders or their valves. Care should be taken not to handle valves with oily hands or oily gloves.
8. Do not expose cylinders to an open flame or to high temperature.
9. Do not expose cylinders to artificially created low temperatures as all cylinders will lose their ductility at some low temperature. Manganese steel type 3A cylinders should not be exposed to temperatures of −20°F or lower.
10. Do not weld or strike an arc on any cylinder, as such operations can dangerously weaken the cylinder wall.
11. Do not attempt to repair a valve or relief device while the cylinder contains gas pressure.
12. Do not handle cylinders with lifting magnets. Do not lift a cylinder by its cap or with slings, ropes, chains, or any other device not specifically designed for lifting cylinders.
13. Keep valves closed except when cylinders are in use or being recharged.
14. Do not use cylinders as clothes hangers.
15. Do not use compressed gas to dust off clothing. This may seriously injure the body or eyes or cause a fire hazard. Above all, do not use oxygen. Clothing saturated with oxygen can ignite from a spark and burn with great intensity.
16. Do not use the gas from a cylinder that is not provided with a legible decal that identifies its contents.
17. When transporting by suitable cart or truck, fasten cylinders securely in upright position so that they will not fall or strike each other.

Cylinder Storage

1. Store cylinders upright, in well-ventilated areas away from flames, sparks, and sources of heat or ignition.

2. Separate oxygen cylinders (empty or full) from fuel-gas cylinders and combustible materials by at least twenty feet or by a barrier at least five feet high having a fire resistance rating of at least one-half hour.
3. Do not store flammable gas cylinders with oxygen or nitrous oxide cylinders or adjacent to oxygen charging facilities.
4. Do not store cylinders near combustibles, elevators, or gangways, or where heavy objects may strike them.
5. Do not store cylinders on the edge of a loading dock or other elevated platform.
6. Do not park trucks with mixed loads adjacent to oxygen or nitrous oxide storage areas except while loading or unloading.
7. Store full and empty cylinders separately and identify by signs to prevent confusion.
8. Do not place cylinders where they might become part of an electric circuit.
9. Protect cylinders stored outdoors from the ground to prevent bottom corrosion. Where extreme temperatures prevail, protect cylinders from the direct rays of the sun.
10. Do not expose cylinders to continuous dampness or store them near corrosive chemicals or fumes.
11. Secured stored cylinders to prevent tipping or falling.

2.6.6 Handling Hazardous Gases

2.6.6.1. Inert Gases
Most inert gases are simple asphyxiants. Their major hazard, which is especially severe when they are used indoors, is displacement of air and the creation of an oxygen-deficient atmosphere.

Air normally contains 21% by volume oxygen. If this concentration is reduced to less than about 12%, humans will become unconscious almost immediately and with no warning. Death will follow quickly if oxygen is not restored.

At oxygen concentrations of 15–16%, an individual becomes mentally incapable of diagnosing the situation and even euphoric, so that he will make no attempt to escape. Whenever oxygen concentrations fall below 19.5%, air-line masks or self-contained breathing apparatus is required.

Details of accident prevention techniques are in CGA P-12 and CGA P-14. CGA SB-2, "Oxygen-Deficient Atmospheres," is a good introduction for personnel.

2.6.6.2. Medical Gases
Medical gases include several of the common compressed gases as well as special mixtures. They often have high standards of purity and always require special precautions in preparation, packaging, handling, and use. These different aspects are covered in a series of codes.

Purity is regulated by the U.S. Pharmacopoeia (USP) and the National Formulary (NF). Use of gases in medical applications is regulated by the FDA. Requirements that must be met for FDA registration are in 21CFR200-299.

The CGA publishes codes that cover

- handling (CGA P-2)
- color coding (CGA C-9)
- piping systems (CGA P-21, piping is also treated in NFPA 99.

2.6.6.3. Toxic Gases

Toxic materials are listed and their permissible exposure limits are given in 29CFR1910 Subpart Z. Each year the ACGIH publishes its "Threshold Limit Values and Biological Exposure Indices." Information on hazardous properties and necessary precautions is on the container labels, in the MSDS, and in specific gas monographs found in Part III of the CGA Handbook.

Training of personnel is particularly important with toxic materials. Some safety precautions and first aid measures may be quite specific. Among the specific precautions is the use of the right gas mask; that is, one equipped with the right adsorbent cartridge. Masks are acceptable only at toxic gas concentrations within the rating of the mask and at oxygen concentrations above the minimum of 19.5–20%.

A self-contained breathing apparatus is an alternative to a gas mask. This must be of a type approved for the particular service by NIOSH, the Bureau of Mines, or other authority.

Some general precautions when using toxic gases follow:

- Store toxic gases outdoors or isolate in separate noncombustible areas.
- Mark storage areas clearly and restrict entry by unauthorized persons.
- Limit the quantity of toxic material stored on site.
- Handle gases outdoors or in areas with forced ventilation (preferably in ventilated hoods).
- Gather emissions from process into appropriate facilities for absorption or destruction of the toxic gas.

2.6.6.4. Reactive Gases

Reactive gases can attack body tissue and other materials. Such gases are identified and precautionary measures to be used in their handling are given in an MSDS or in one of the specific gas monographs in Part III of the CGA Handbook. Personal protective equipment includes clothing, shields, and breathing apparatus. The remarks made in the preceding section on gas masks and breathing apparatus apply here. Other precautions include:

- Avoid contact of acid or alkaline gas with skin or eyes (use goggles or face shields).

- Handle reactive gases only in well-ventilated areas equipped with safety showers and eyewash fountains.
- Be sure that first-aid and medical personnel, on and off site, have accurate information on hazards, symptoms, and treatment.
- Limit quantities of reactive gases stored on site, and store outdoors or in well-ventilated areas.

2.6.6.5. Oxygen

Gases with high oxygen concentrations react exothermically with many substances. If not controlled, these reactions can lead to explosions or detonations. Similar problems can arise with other oxidizing gases such as halogens, NF_3, and N_2O. Recommended precautions include:

- Clean all equipment to be used in oxygen service to remove all traces of hydrocarbon materials.
- Do not allow combustible substances to contact cylinders, valves, regulators, gases, hoses, or fittings used for oxidizing gases.
- Do not lubricate valves, regulators, gauges, or fittings with a combustible substance.
- Do not handle cylinders with oily hands or gloves.
- Do not store oxidizing gases near combustible material or sources of ignition.
- Inspect cylinders regularly to guard against loss of strength from oxidation of wall (see CGA C-6).
- Remove clothing immediately when exposed to liquid oxygen or oxygen-rich atmosphere, and air out before reuse (see CGA P-14).
- Exclude combustible materials from areas in which liquid oxygen may be spilled during transfer. This includes the situation in which combustibles may be introduced accidentally and then held or accumulated (e.g., oil drippings on porous surface such as gravel).

A large body of published information is available on oxygen and its safe handling. Suppliers of the gas are an excellent source of information. The use of oxygen places definite restrictions on the design of facilities. The National Electrical Code (NFPA 70) must be satisfied. NFPA 50 covers bulk oxygen systems on consumers' sites, and NFPA 43C covers the storage of gaseous oxidizing materials. The requirements for cleaning all components of an oxygen-handling system can become very complex. This is covered in CGA G-4.1 and is the subject of Chapter 10 of the CGA Handbook.

2.6.6.6. Acetylene

There are a few large-scale synthetic applications of acetylene. A smaller amount is used for welding, cutting, etc., but these are much more frequent applications. Most plants probably will receive and handle acetylene only in cylinders.

The principal hazards of acetylene are the flammability and instability of the gas. It is an asphyxiant, but only above its LFL. Piping should follow ANSI B31.3. Metals that can form unstable acetylides, of which copper is the common example, must be avoided. NFPA 51, ANSI Z49.1, and CGA SB-8 all cover welding and cutting installations. Some of the precautions that apply to acetylene follow:

- Storage
  - —Do not store near oxygen cylinders.
  - —Limit amount stored in a building to 2500 ft.[3]
  - —Store in vertical position (prevents loss of solvent and spread of flame from failed plug).
  - —Protect against bumps from workers, vehicles, etc.
- Handling cylinders
  - —Secure cylinders in use by nesting, chaining, etc.
  - —Do not alter or repair cylinders, valves, or protective devices.
  - —Keep away from flames and sparks.
  - —Do not refill cylinders or transfer gas from one cylinder to another.
  - —Do not mix gases in a cylinder.
  - —As much as possible, keep electrical and oxyacetylene welding equipment separated.
  - —Use only warm water to thaw ice in fittings.
- Moving cylinders
  - —Protect against dropping and slamming.
  - —Transport with regulator off, valve closed, and valve cap on.
  - —Do not use lifting devices that transfer part of the load to the cylinder itself (e.g., rope, chain, sling); support cylinder on cradle.
  - —When using hand truck, keep cylinder upright and chain securely.
- Use
  - —Do not use at pressures above 15 psig.
  - —Do not use shutoff valve on supply line unless cylinder is equipped with a regulator.
  - —Open and close valves slowly; restrict valve opening to prevent rapid accidental discharge.
  - —Use regulators approved for acetylene service (danger zone marked in red).
  - —With wrench-operated valves, use tool approved by manufacturer; leave wrench on valve at all times when in use.
  - —Keep all connections and hoses leaktight.
  - —Bleed pressure from valve connections and regulators when shut off.
  - —Use soap to test lines and connections for leaks.
  - —Follow NFPA 51 recommendations for fire protection.
  - —When a valve or fusible plug is leaking, tag the cylinder, move it to an open area, place warning signs in the area, and consult the supplier; open valve slightly to allow slow escape of acetylene.
  - —Consult CGA SB-4 for advice in fire situations.

- Shipping
  —Check pressure against maximum allowed by 49CFR173.303(b).
  —Verify that cylinder meets specifications referred to above.
  —Follow CGA C-13 when requalifying cylinders.
  —Do not ship or allow shipment of cylinders in poorly ventilated
    containers or those in which heat buildup can occur.
  —Be sure that cylinders are properly secured when shipped.

2.6.7. Cryogenic Liquids

2.6.7.1. Hazards
Special hazards associated with cryogenic liquids include:

- extreme cold which can freeze body tissue on contact and embrittle
  normal carbon steel, plastics, and rubbers (temperatures encountered
  in sprays or fogs are often considerably lower than the normal boiling
  point of a material)
- high pressure that can develop from heat leaks into a storage container
- displacement of air by asphyxiating gas upon leakage of liquid from
  system
- fire or explosion caused by escape of flammable material or oxygen
- splattering of cold liquid because of extremely rapid vaporization of
  escaping liquid

Special precautions include the use of protective clothing, including eye
or face protection and heavy leather gloves. Transfer operations, especially
those involving open containers such as Dewars, should be carried out slowly
and carefully to control boiling and splashing of the liquid. Ventilation must
be adequate to handle any gas evolved.

In system design, materials of construction must be compatible with the
liquids handled at the low temperatures to be encountered. Apparatus must
have proper pressure relief. Where liquid can be trapped between valves, the
possibility of high pressure due to partial vaporization or thermal expansion
must be considered. Again, pressure relief devices may be called for.

Other details may be found in CGA P-12.

2.6.7.2. Specific Precautions
Because of their different physical and chemical properties, the various cryogenic liquids have their own peculiar requirements, some of which are discussed below.

Oxygen
Because of its explosive reactivity with many substances, oxygen presents unique
problems. Any system that will contain oxygen must be cleaned and kept free of

flammable materials. Oils and greases used for cutting or applied for protection are frequently overlooked examples. Cleaning is discussed in CGA G-4.1.

Keeping reactive materials at a distance from a bulk oxygen system also provides some safety. Standards are published in NFPA 50.

Other points to recognize are:

- Liquid oxygen spilled onto clothing poses a serious hazard. Such clothing must be removed at once.
- Oxygen can not be effectively blanketed by fire extinguishing agents. Water sprays are usually the best protection.
- Contact of liquid oxygen with a combustible material may not immediately cause a fire, because of the combination of low temperature and lack of an ignition source. It is important to keep ignition sources away from a spill of liquid oxygen until it dissipates, and to recognize that equipment moved across a combustible surface (e.g., asphalt) or even foot traffic can cause ignition.

Other Cryogenics

Hydrogen, helium, and LNG are other cryogenics that require specific precautions. These are spelled out in the publications of the CGA and the NFPA and summarized briefly below.

Liquid	Code	Precautions
LNG	NFPA 59A	Positive-pressure storage
		Labeling
		Electrical classification
Helium	CGA P-12	Positive-pressure storage
		Prevention of backflow and
		diffusion of air into system
		(to avoid solidification)
Hydrogen	NFPA 50B	Similar to LNG and helium
		Outdoors
		Equipment designed for H_2 service

DOT 4L Cylinders

DOT 4L cryogenic containers are double-walled, vacuum, multilayer insulated cylinders designed for product withdrawal in either gaseous or liquid form. Such containers are generally used for liquefied argon, helium, hydrogen, neon, nitrogen, and oxygen.

These cylinders typically comprise a stainless-steel inner container encased within an outer steel vacuum shell. To protect the inner container from overpressurization, the unit includes a CG-7 pressure relief valve (see Table 2-11). Such containers are further protected by a CG-1 rupture disk as a secondary relief device. The cylinder may also contain an internal vaporizer, which converts cold liquid to warm gas for withdrawal.

2.7. WASTE HANDLING

The handling of plant waste can raise a variety of health and safety concerns. These wastes include not only process materials but also empty containers, the refuse from maintenance activities, and even process equipment that has been removed from the plant. The owner must establish a plan that covers all these wastes.

2.7.1. Waste Disposal Plan

The owner must:

- identify the volumes and types of waste produced
  - —vapor or gas
  - —liquid
  - —solid, including empty containers and scrapped equipment, piping, and fittings
- establish a plan for minimizing waste by
  - —limiting the use of raw materials (e.g., by using higher grades of materials that produce less waste)
  - —recovering and recycling waste material to the process
  - —establishing a use or market for waste products
- establish a disposal plan for handling and treatment of the remaining waste by
  - —defining a procedure for handling each substance
  - —providing standby means for containing hazards and for maintaining critical communications
  - —providing warning signals or signs
  - —controlling access to area
- determine the need for waste storage facilities
  - —space allocation
  - —segregation of hazardous waste
  - —packaging and labeling
  - —monitoring
  - —special ventilation
  - —special environmental control
  - —compliance with regulations
- establish personal protective equipment requirements for handling waste [33, 34].
- satisfy all environmental regulations

Some of the necessary activities should be undertaken only with permits (Section 7.1.4). Treatment methods might include

Recovery	Vapor reclamation
	Filtration
	Electrostatic precipitation
	Wet scrubbing
	Adsorption
Thermal treatment	Incineration
	Detonation
	Pyrolysis
	Open burning
Chemical treatment	Acid/base neutralization
	Ion exchange
	Oxidation/reduction
	Precipitation/clarification
Biological treatment	Microorganism addition
	Activated sludge
	Aerobic digestion
	Anaerobic digestion
	Landfarming
Land disposal	Secure chemical landfill
	Deepwell injection
	Surface impoundment

For waste drums and other containers, the plan should include procedures for

- handling bulging, leaking, and deteriorated containers, as well as those holding material that is
  —shock-sensitive or explosive
  — reactive
  — chemically toxic
  —biologically toxic or genetically altered
  —radioactive
- opening containers
- sampling
- analyzing contents
- transferring contents to another container
- decontamination
- labeling and storage
- shipping for disposal off site

2.7.2. Release Reporting

Several different sets of regulations require a manufacturer to report certain releases to the environment. Toxic materials are covered under right-to-know legislation. 40CFR372 gives a specific list of chemicals and chemical categories. It also defines

- facilities covered by the rule
- thresholds for reporting
- report requirements
- schedule for submission
- exemptions
- supplier requirements
- recordkeeping

Each facility is responsible for understanding and meeting its obligations under this rule.

EPA also has published rules (40CFR300, 302, and 355) which require that continuous releases of hazardous substances be reported. First, the person in charge of a facility must satisfy the requirements of Section 103 of CERCLA by initial notification of releases that exceed the reportable quantity for any material on the CERCLA list of hazardous substances.

The rules mentioned above allow less frequent reporting of releases that are continuous and stable in quantity and rate. The definition of "continuous" encompasses releases that are continuous during regularly occurring batch processes or during operating hours. When a given release qualifies under these rules, the reporting obligation is reduced to

- immediate reports of any statistically significant increases
- an annual report

When a release is determined to be a statistically significant increase above background level, it must be reported as set out in 40CFR302.6. It is the reporter's responsibility to convince EPA of the adequacy of the statistical test that is used. 40CFR302.8 Appendix C contains a simple nonparametric test that will be accepted. This involves simply reporting any release that is greater than each of the previous nineteen. The annual report should include

- identity of hazardous substance
- total amount released
- basis for claiming "continuity" and "stability of quantity and rate"
- dates on which release exceeded reportable quantity
- sizes of mean release and largest release
- number of statistically significant increases reported
- nature of significance test

2.7.3. Scrap and Salvage

Hazards also result from maintenance and decommissioning of process equipment. Some of these are:

- handling of refuse from maintenance of operating equipment
- decommissioning of inoperative equipment that will remain in place

- decommissioning and dismantling of equipment that will be shipped off site for renovation or disposal

Many of these hazards occur while handling the scrap and salvage materials produced by the following activities:

- cuts and scratches from handling sharp and pointed scrap materials
- respiratory problems resulting from the production of dust during dismantling and cleanup
- eye and respiratory problems resulting from splashing or production of toxic or irritating vapor during dismantling

Decommissioning of inoperative equipment will make it nonhazardous to plant personnel. The equipment must be decommissioned in order to:

- remove it from the process train where it could adversely affect safety
- remove all hazardous residues from it
- determine whether the equipment should be abandoned in place or dismantled for disposal

Essential features of an equipment cleaning/decontamination process are:

- method effective for removing contaminants
- decontamination materials compatible with materials of construction of equipment and with hazardous substances present in system
- decontamination materials and process hazards offset by protective measures
- safe and acceptable means of disposal of used decontamination solution

2.7.4. Vessel Decommissioning

Process and storage tanks that are no longer useful are usually the most difficult to decontaminate and remove. The steps in vessel decommissioning are as follows:

- Identify the contents of the vessel.
- Isolate the vessel from piping and other process equipment.
- Empty and decontaminate the vessel. If the vessel contains volatile vapors, vent excess pressure; however, protect personnel from possible rapid releases with deflecting shields.
- Issue permits and follow safe procedures for vessel entry and work in confined spaces (see Section 7.1.4).
- Proceed to decommission or dismantle the vessel.
- Provide for proper shipment of dismantled vessel.

2.7.5. Waste Containers

2.7.5.1. Disposal of Used Drums

The steps in handling waste drums are summarized below and charted in Figure 2-22.

- Determine the drum type: metal, PVC-lined, etc.
- Determine the contents of the drum and its hazards
  —explosive or shock-sensitive
  —corrosive
  —poisonous
  —biologically hazardous
  —radioactive
- Examine the drum to determine:
  —is it in good condition?
  —can it be opened correctly?
  —is it under pressure and bulging?
  —is it deteriorated and leaking?
- Determine whether the contents should be transferred in place or after moving to a remote location.
- Determine need and methods for decontamination.
- Determine proper means of disposal or shipment off-site.

If a drum is swollen or bulging, it should be handled very carefully. The cause of the swelling should be determined and procedures modified to prevent recurrences. The following steps to reduce the swelling should be performed slowly:

- Determine whether cooling the drum will reduce the pressure causing the swelling.
- Determine whether the bung can be gradually opened to reduce the pressure causing the swelling.
- Determine the proper method for opening the drum:
  —Use remote-controlled devices such as those shown in Figures 2-23 and 2-24.
  —Attempt to open the drum manually and slowly using the correct bung wrench. Do not use picks, chisels, or other tools that can shock the contents of the drum or cause an uncontrolled release. If the drum is to be opened manually, minimize personnel exposure by covering or enclosing the drum with material that can contain a release of the contents.
- Determine the proper method for transferring the contents to a safer container.
- Determine the proper method of staging and disposing of the drum.

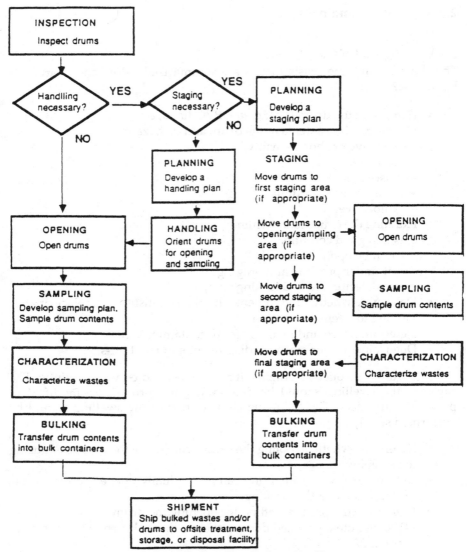

FIGURE 2-22. Flow Chart for Drum Handling.

2.7.5.2. Disposal of Used Cylinders

It frequently will be necessary for the user to arrange the safe disposal of a compressed gas cylinder. With most types, disposal will be required because of the discovery of a leak or some other flaw. Certain types, moreover, are officially nonrefillable (e.g., DOT 39, 40, and 41). For such cylinders, the

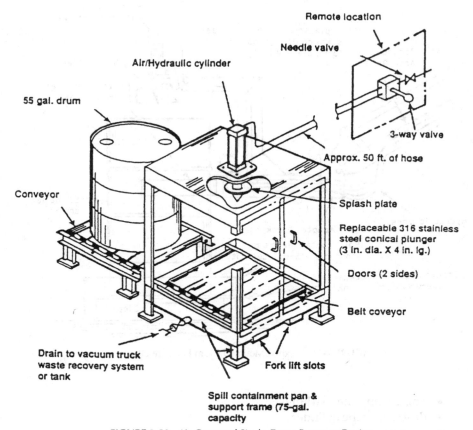

FIGURE 2-23. Air-Operated Single-Drum Puncture Device.

supplier should give instructions and procedures for return or safe disposal. These instructions should be followed precisely.

When a cylinder is damaged in some way, it and its contents should be disposed of by qualified personnel. Potential hazards are the pressure within the cylinder, the flammability of the contents, the possible formation of explosive mixtures, and the release of toxic or reactive materials.

When any of the necessary activities is beyond the capability of the organization, outside help should be brought in. Generally, the procedure includes

- identifying the contents
- emptying and purging the cylinder (see below on safe disposal of contents)
- removing the valve and other connections

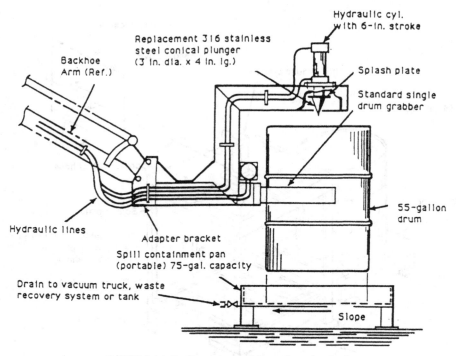

FIGURE 2-24. Backhoe-Mounted Drum Puncture Device.

- obliterating the markings
- destroying the cylinder
- properly discarding the remnants.

Specifically, the user should

- locate the supplier of the cylinder and, if possible, return it
- if the supplier can not be identified, find nearby manufacturers or distributors of the gas and request that they dispose of the gas
- if shipping is not feasible or the gas can not be handled as above, obtain instructions for disposal from the same sources and request their help in supervision

Other instructions are in CGA C-2.

2.7.5.3. Disposal of Contents

The appropriate method for disposal of the contents of a cylinder can be quite specific. A relatively simple case is the disposal of oxygen or inert gases. These usually can be vented to the atmosphere, at a controlled rate and with good ventilation. Flammable materials can be destroyed by controlled burning, if proper care is taken in handling the combustion products. Toxic and noxious materials require specific approaches. Some materials can be collected for use or later disposal. Others may be absorbed or chemically destroyed. Phosgene, for example, is destroyed by contact with caustic soda.

Such treatment should be planned for, and facilities should be in place for any operating emergencies. Again, the gas supplier or industry association should be a good source of information on methods of safe disposal.

When a cylinder valve is faulty, the cylinder should be secured physically in case of rapid release. A valve with small aperture should be connected to the cylinder to allow control of the venting rate.

Pressure relief devices are to be handled only by experts.

REFERENCES

1. Lewis, R. J., Sr., *Sax's Dangerous Properties of Industrial Materials*, 8th ed., 3 vols., Van Nostrand Reinhold, New York, 1992.
2. Budavari, S., ed. *The Merck Index*, 11th ed., Merck and Co., Rahway, NJ, 1989.
3. *NIOSH Pocket Guide to Chemical Hazards*, Department of Health and Human Services, GPO, Washington, 1990.
4. *Federal Register*, *48*, 53280–53348 (25 November 1983) and *52*, 31852–31886 (24 August 1987).
5. 29CFR1910.1200, Hazard Communication, GPO, Washington.
6. National Fire Protection Association, *Recommended Systems for the Identification of the Fire Hazards of Material*, No. 704, NFPA, Quincy, MA, 1985.
7. Title 49, Code of Federal Regulations, Part 172, Subpart E, GPO, Washington.
8. Department of Transportation, *ERG90—1990 Emergency Response Guidebook*, Edition DOT P 5800.5, GPO, Washington, 1990.
9. *Plant Engineering* 45, No. 10, 6 June 1991, p. 10.
10. Crowl, D. A., and J. F. Louvar, *Chemical Process Safety: Fundamentals with Applications*, Prentice-Hall, Englewood Cliffs, NJ, 1990, p. 170.
11. National Electrical Code, Articles 500 through 503, National Fire Protection Association, Quincy, MA.
12. NEMA 250-1985, *Enclosures for Electrical Equipment (1000 Volts maximum)*, National Electrical Manufacturers Association, Washington.
13. Hathaway, G. J., et al., *Chemical Hazards of the Workplace*, 3d ed., Van Nostrand Reinhold, New York, 1991.
14. Dow Chemical Company, *Chemical Exposure Index Guide*, American Institute of Chemical Engineers, New York, 1994.
15. Hodge, H. C., and J. H. Sterner, American Industrial Hygiene Quarterly, *10*:93, "Tabulation of toxicity classes," 1949.
16. Young, J. A., *Improving Safety in the Chemical Laboratory*, John Wiley & Sons, New York, 1991, Section 9.10 Glossary, pp. 143–146.
17. DHHS Publication No. (CDC) 84-8395, *Biosafety in Microbiological and Biomedical Laboratories*, U.S. Department of Health and Human Services, Public Health Serv-

ice, Center for Disease Control and National Institutes of Health, GPO, Washington, 1984.

18. National Safety Council, *Accident Prevention Manual for Industrial Operations*, volumes entitled "Administration and Programs" and "Engineering and Technology," 10th ed., NSC, Chicago, 1992.

19. Cralley, L. J., and L. V. Cralley, eds. *Industrial Hygiene Aspects of Plant Operations, Volume 3—Equipment Layout and Building Design*, Macmillan, New York, 1986.

20. Apple, J. M., *Plant Layout and Material Handling*, 3d ed., John Wiley & Sons, New York, 1977.

21. Mecklenburgh, J. C., ed. *Process Plant Layout*, John Wiley & Sons, New York, 1985.

22 Benedetti, R. P., ed. *Flammable and Combustible Liquids Code Handbook*, 4th ed., National Fire Protection Association, Quincy, MA, 1991.

23. Dow Chemical Company, *Dow's Fire and Explosion Index Hazard Classification Guide*, 7th ed., American Institute of Chemical Engineers, New York, 1994.

24. Kletz, T., *Plant Design for Safety, A User-Friendly Approach*, Hemisphere, New York, 1991.

25. Geyer, W., Bringing Storage Tanks to the Surface. *Chem. Eng.* 99, No. 7, July 1992, pp. 94–102.

26. Hart Environmental Management Corp., *Underground Storage Tank Management: A Practical Guide*, 2d ed., Government Institutes, Inc., Rockville, MD, 1987.

27. Shamlou, P. A., *Handling of Bulk Solids*, Butterworths, London, 1988.

28. Muir, D. M., ed. *A User Guide to Dust and Fume Control*, I. Chem. E., Rugby, UK, 1985.

29. Compressed Gas Association Inc., *Handbook of Compressed Gases*, 3d ed., Van Nostrand Reinhold, New York, 1990.

30. Tribolet, R. O., Compressed Gas Association, personal communication to T. F. O'Brien, 16 April, 1993.

31. O'Brien, T. F., and I. F. White, *Modern Chlor-Alkali Technology*, Vol. 5, pp. 239–256, Elsevier Applied Science, London, 1992.

32. Senesky, J., *Plant Engineering* 33, No. 8, pp. 289–292, and No. 10, pp. 143–148 (1979).

33. Olishifski, J. B., ed. *Fundamentals of Industrial Hygiene*, 2d ed., National Safety Council, Chicago, 1979, p. 274.

34. Department of Health and Human Services, NIOSH Publication No. 85-115, *Occupational Safety and Health Guidance Manual for Hazardous Waste Site Activities*, DHHS, GPO, Washington, 1985.

3

PROCESS EQUIPMENT AND PROCEDURES

This chapter covers safety considerations in design, installation, operation, and maintenance of selected process equipment and auxiliary systems. The first section (3.1) briefly discusses the impact of material selection on process equipment safety. Section 3.2 introduces the subject of corrosion and its control. The following more specific topics are then covered:

- small containers (3.3)
- piping (3.4)
- transfer hoses (3.5)
- pumps (3.6)
- fans and compressors (3.7)
- motors and turbines (3.8)
- filters (3.9)
- centrifuges (3.10)
- drying and particle size reduction equipment (3.11)
- instruments and controls (3.12)

The vital subjects of pressure relief and the use of safety valves and other protective devices would normally be considered in the context of pressure vessel safety. Since this chapter does not discuss vessels directly, some considerations on pressure relief and a list of applicable standards are included in Section 3.4.2 on piping.

Each of the types of equipment listed above has its own specific requirements. A group of common considerations, however, applies to all of the above topics.

Design
- The application must be well defined.
- The nature of each material to be processed must also be properly defined. There should be a list of all important components, with concentrations and their variability.

- Safe operating limits must be defined.
- The equipment must match the original specifications for the application and be in accord with design and vendor's drawings.
- Materials of construction and seals must be appropriate for the application.

Installation

- The equipment as received must be in accord with purchase specifications and drawings.
- The location must be prepared for installation; necessary facilities and personnel must be available to move and install the equipment.
- The mounting must be secure and in accord with specifications.
- The equipment must be oriented and connected properly.

Operation

- The equipment must be operated within design conditions.
- Operating instructions must be available and scrupulously observed.
- There must be no overloading or surging in operating conditions that might jeopardize the safety of the operation.

Maintenance

- Clear instructions, such as a vendor's recommendations, must be available.
- Personnel must be trained in maintenance activities and the use of special tools.
- Maintenance schedules must be followed, and records must be up to date.
- A preventive maintenance program must be developed, involving both operating and maintenance personnel. Tests, inspections, and replacement of parts or fluids should be to schedule. All deviations and unexplained changes in operation should be reported and promptly investigated.

3.1. MATERIALS OF CONSTRUCTION

Plant safety depends on proper selection of materials for process piping and equipment. Misapplication of a material or poor design of components can cause such problems as:

- short service life
- catastrophic failure during operation
- static electricity accumulation which in the presence of flammable substances increases the potential for fire or explosion

Problems can arise from the mismatch of different materials that would not fail had they not been joined together. Failure is often caused by galvanic corrosion, which is discussed in Section 3.2.

Familiarity with the mechanical properties of materials is essential to their proper selection. Some important properties and their impact on process safety are:

- *Brittleness:* the tendency to fracture at low stress without warning. Plastics, FRP, glass, and cast iron must operate well below the stress levels that cause fracture and must also be protected from damage by impact. Brittleness tends to increase at low temperature and may exhibit a sort of transition temperature below which it increases greatly.
- *Creep:* the plastic deformation of material under load. Such deformation will eventually cause a material to fracture. Most ferrous alloys are subject to creep above 800°F. Lead, if not properly supported, will creep at room temperature. Most plastics, almost by definition, will creep.
- *Ductility:* the ability to deform permanently under tensile stress without breaking. This is a desirable property for materials used in pressure piping and vessels.
- *Fatigue:* the progressive fracturing of a material by cyclic stresses. Ductile metals have a higher resistance to fatigue than hard metals, ceramics, and most plastics. Materials with high resistance to fatigue should be used for components of positive displacement pumps and compressors. Steels, magnesium, and nickel alloys have good fatigue resistance; nickel alloys have better fatigue resistance at elevated temperatures than steels. Copper and titanium alloys have intermediate resistance, while those of aluminum are relatively poor. Stress raisers will also decrease fatigue resistance.
- *Hardness:* the opposite of ductility. Hardness is characteristic of tool steels and is a desirable property for materials used in grinding and milling.
- *Impact Strength:* the ability of a material to withstand shock loading without breaking. This property is found in most fibrous materials, such as wood and wrought iron.
- *Malleability:* the ability of a material to be permanently deformed with ease, as in forging, rolling, or extruding. Aluminum, copper, and magnesium have outstanding malleability and can impart this characteristic to their alloys.
- *Toughness:* the ability to withstand loads, including shock, without breaking. This is true of abrasion resistant materials, such as the Hastelloys.
- *Electrical and Thermal Conductivity:* these properties are strongly correlated, high thermal conductivity usually indicating high electrical conductivity. Materials with high conductivity should be used in situations that require fast response to temperature swings or rapid dissipation of static electricity.

3.1.1. Material Selection

Proper selection of materials for process applications requires the following:

- clearly defined process
  —all chemicals, with their properties, listed
  —types of equipment, with their functions and operating conditions, listed
- materials selected with proper mechanical and chemical properties, based on:
  —past experience with similar equipment in similar applications
  —existing plant piping and equipment standards
  —vendor-supplied information on documented past experience
  —pilot study or testing of material applicability and reliability

The quality of the materials must be confirmed by existing codes and standards (see Table 3-1 for sources) or testing and analysis of materials.

TABLE 3-1
Sources of Materials Standards

Abbreviation	Organization	Scope of Concern
ASTM	American Society for Testing and Materials 1916 Race Street Philadelphia, PA 19103	Provides testing procedures for various process materials to ensure quality and uniformity as defined by the standard.
ASME	American Society of Mechanical Engineers 345 East 47th Street New York, NY 10017	Provides list of recommended metals, welding rods, and electrodes for use on pressure vessels.
AISI	American Iron and Steel Institute 1000 16th St., NW Washington, DC 20036	Provides a standard description of carbon steels and alloy steels.
SAE	Society of Automotive Engineers 400 Commonwealth Drive Warrendale, PA 15096	Provides a four-digit code for identifying carbon, alloy, stainless, and tool steels. These standards were developed in conjunction with ANSI and are referred to as the AISI–SAE system.
CDA	Copper Development Association 405 Lexington Avenue New York, NY 10174	Provides a numerical coding system for identifying copper alloys.
ANSI	American National Standards Institute 1430 Broadway New York, NY 10008	Provides safety and design criteria and test procedures for various mechanical components and systems. This includes piping standards with a list of recommended materials.

3.1.2. Material Application

After the proper material has been selected, equipment must be properly assembled and operated to ensure reasonably long service life. Some points to consider to avoid mechanical failures are:

- Reduce or eliminate stress risers in equipment components (see Figure 3-1).
- Limit the friction on moving parts by proper alignment, good lubrication and adequate clearance.
- Eliminate unnecessary vibration of rotary components, which can cause galling and fretting.
- Use high-strength materials wherever fatigue may be a problem.
- Use smooth finishes to reduce friction and buildup of contamination and to increase fatigue resistance.
- Confirm that material properties match the application, for example:
  Low friction   Fluoroplastics, MoS_2
  Strength at high temperature   Inconels, Hastelloys
  Seawater corrosion resistance   Copper, nickel, titanium alloys
  Creep resistance   Steels and nickel alloys

3.2. CORROSION

3.2.1. Types of Corrosion

Corrosion is the degradation of a material caused by exposure to an environment. All material will degrade with time, but the degradation process can be retarded significantly by careful consideration of the following factors:

- selection of materials of construction
- mechanical design of the equipment
- nature of the operating environment

These three factors must be considered together in order to understand how materials corrode. In most cases, the last factor cannot be changed, the environment being defined by process requirements. However, there are some methods that can be used to mitigate corrosion problems:

- eliminating unnecessary moisture in such gases as halogens, hydrogen fluoride, hydrogen chloride and ammonia
- increasing the temperature of a gas above its dew point
- adjusting the pH of a waste before it enters a treatment facility

Corrosion is not always a uniform wearing away of material. Nonuniform corrosion phenomena can be classified as macroscopic or microscopic [1–6]. Macroscopic corrosion mechanisms include:

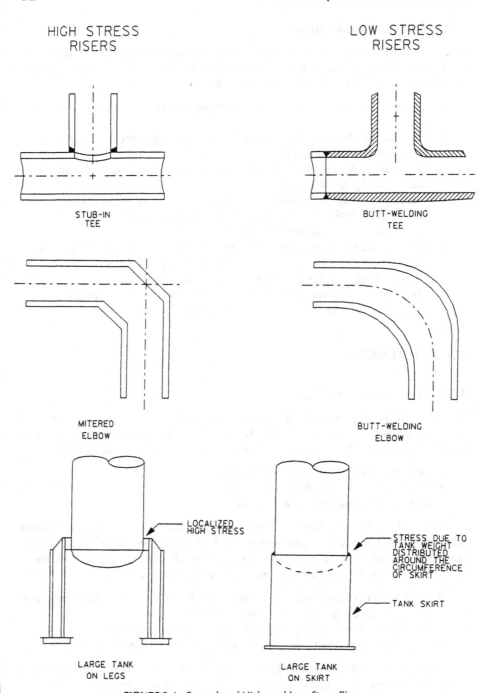

FIGURE 3-1. Examples of High- and Low-Stress Risers.

- *Galvanic:* Corrosion that is electrochemical in nature. Galvanic corrosion is caused by placing two dissimilar metals in contact with each other, directly or through a conductive fluid. The rate of this type of corrosion increases with the distance between the metals in the galvanic series (see Table 3-2). The further apart the metals, the more rapidly the anodic material will corrode. Variations of galvanic corrosion include

TABLE 3-2
Galvanic Series of Metals and Alloys in Seawater

ANODIC (Active) Least Noble Metals [Most susceptible to corrosion]	Magnesium and magnesium alloys Zinc Galvanized steel, galvanized wrought iron Aluminum and aluminum alloys Cadmium Low-carbon steel Alloy carbon steel Wrought iron Cast iron Stainless steel, Type 410 (active) Stainless steel, Type 430 (active) Stainless steel, Type 304 (active) Stainless steel, Type 316 (active) Lead Muntz metal Yellow brass Admiralty brass Aluminum brass Red brass Copper Silicon bronze 90/10 copper–nickel 70/30 copper–nickel Nickel Inconel Silver Stainless steel, Type 410 (passive) Stainless steel, Type 430 (passive) Stainless steel, Type 304 (passive) Stainless steel, Type 316 (passive) Alloy 20 Monel
CATHODIC (Passive) Most Noble Metals [Will cause less noble material to corrode]	Hastelloy Titanium Graphite Gold Platinum

thermogalvanic corrosion, caused by large temperature gradients, and stray current corrosion, caused by improper grounding of electrical equipment.

- *Erosion:* Corrosion caused by the impingement of a corrosive fluid or fluid containing abrasive particles on a component, removing or preventing the formation of a protective coating. Corrosion resistance of galvanized and passivated metals can effectively be eliminated by erosion.
- *Pitting:* Corrosion in the form of crevices on a surface.
- *Fatigue:* The combination of mechanical stress cycling and exposure to a corrosive medium.

Microscopic corrosion mechanisms include:

- *Intergranular:* Corrosion that preferentially attacks the grain boundaries of a metal. Although it cannot be seen, it can cause a catastrophic failure of the metal. Stainless steels are commonly attacked in this manner, particularly those with high carbon content and improper heat treatment of welds.
- *Stress cracking:* Corrosion in the form of crack formation around a point of high stress concentration exposed to a corrosive environment.
- *Dealloying:* Corrosion in which an alloying element is preferentially attacked and dissolved.
- *Hydrogen embrittlement:* Cracking or blistering caused by hydrogen entrapment in the metal. This often is caused by improper welding and may be found in acid pickling operations and high-temperature reactions where atomic hydrogen is present.

Microscopic corrosion is more difficult to detect by routine inspection or thickness measurement and so is more likely to result in catastrophic failure.

In order to prevent these various types of corrosion, the following should be considered:

- Avoid materials that are susceptible to hydrogen embrittlement [6, 7].
- Eliminate stress risers in equipment components to reduce stress corrosion cracking.
- Use materials that are not susceptible to stress corrosion cracking.
- Reduce the potential for erosion by reducing velocity of impingement on surfaces such as vessel walls or piping tees and elbows.
- Insulate legs on hot tanks to prevent thermogalvanic corrosion at welds.
- When two metals far apart in the galvanic series (Table 3-2) are placed together, take one of the following steps:
  —insert insulator between the two metals.
  —attach conducting wire to each piece by a clamp or bolted flange, using the bolt or flange as a sacrificial anode.
  —make anodic (less noble) metal part larger than cathode in order to slow the corrosion rate.

- Properly ground all electrical equipment to prevent galvanic corrosion from stray currents (see Figure 3-2).
- Eliminate crevices wherever possible. Crevices allow corrodent materials to concentrate (see Figures 3-3 and 3-4).
- Drain corrosive materials completely from vessels to prevent accumulation of concentrated corrodent in the vessel.
- If a buried pipe is corroding, determine whether:
  —corrosion is external or internal
  —pipe is properly grounded (see Figure 3-2)
  —cathodic protection is necessary
- Electrically isolate pipelines from facilities.

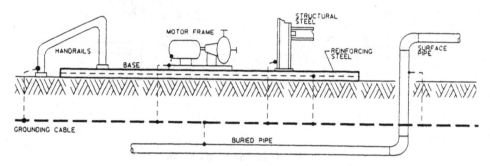

FIGURE 3-2. Grounding of Equipment and Piping.

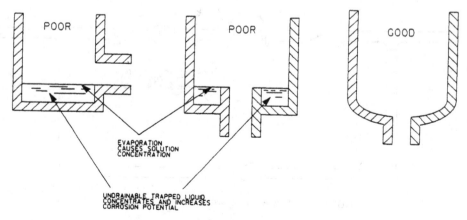

FIGURE 3-3. Examples of Good and Bad Drainage.

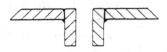

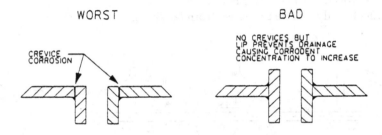

FIGURE 3-4. Elimination of Crevices.

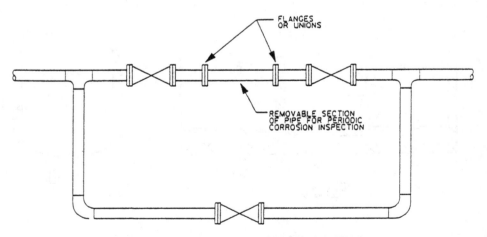

FIGURE 3-5. Pipe Assembly with Corrosion Test Pipe.

3.2.2. Sources of Corrosion Information

There are several ways to determine the potential for corrosion:

- Examine existing data on similar processes for possible causes and cures.
- Consult the Corrosion Data Survey published by the National Association of Corrosion Engineers (NACE).
- Consult equipment suppliers for documented successful applications.
- Insert oxidation–reduction potential (ORP) probes throughout a test system.
- Test in laboratory, using actual or simulated process conditions with the materials under consideration.
- Include a pipe assembly such as shown in Figure 3-5 to periodically check the integrity of the system.
- Run tests on various materials under comparable conditions.

Troubleshooting and corrosion prevention are covered in many standard texts [1, 2, 4–8].

3.3. SMALL CONTAINERS

A "small container" is any commercially available package whose capacity is no more than fifty-five gallons. It may be of any of a wide variety of materials of construction. The category includes drums, cans, bags, and fiber cartons, but excludes sample containers and unique or specially manufactured items. Cylinders holding compressed gases are treated in Section 2.6. This section considers both raw material containers as received and product containers for shipment.

3.3.1. Container Specification

In choosing a container, one must be aware of agency requirements (such as those of DOT in CFR Titles 21 and 49) and of other requirements specific to the contents. Most container manufacturers can help to answer the following questions:

- Does the material of construction give adequate protection against corrosion?
- Is there adequate allowance in the loaded container for thermal expansion of the contents?
- Will the container withstand the vapor pressure of the contents at the highest expected temperature?
- Is the container adequate for the proposed level of stacking?

A special liner will often be used to protect the container from corrosion or the stored material from contamination. However, the liner itself must not contaminate or react with the material being stored. The possible effects of liner failure must be considered:

- failure of the package to contain the material
- unusual or hazardous reaction
- loss of ability of the container to withstand stacking or internal pressure

3.3.2. Manufacturer's Quality Control

To ensure that purchased containers meet the user's specifications consistently and reliably, the quality control program of the container manufacturer is important. When containers are used in critical applications, the purchaser should audit the manufacturer's performance regularly. The audit should cover:

- adequacy of quality control steps in the manufacturing process
- manufacturer's checks on:
  —quality control of container raw materials
  —compliance with DOT tests
  —compliance with purchaser's specified tests or special requirements
- sealing of container openings before shipment to prevent entry of foreign material
- proper support of containers during shipment
- containers loaded so as to allow safe and convenient unloading

3.3.3. Receiving

Many considerations affect the safe handling of new or reconditioned containers, whether empty or full. When a container is loaded, many hazards are aggravated by the increased bulk or weight; in addition, other hazards may be presented by the contents. This section assumes that containers are filled.

In whatever form, the containers are probably received in a vehicle or similar mobile device. The following checklist addresses both the vehicle itself and the containers:

- Provide adequate protection against pinching or crushing of personnel moving the shipping vehicle into place.
- Ensure that trailers are properly chocked.
- Place trailer jacks under the front end of a trailer if the tractor is not attached.
- Ensure that personnel understand how to open the trailer safely if load is unstable.
- Provide adequate walkways in the trailer if containers are stacked.

- Stack containers properly to prevent instability.
- Ensure that containers are securely supported.
- Check condition of skids.
- Make personnel aware of the hazards specific to the unloading process, such as:
  —back injury
  —pinching hazard
  —falling hazard (unstable load).
- Inspect containers (statistical survey) to ensure that purchase specifications are met, including:
  —internal inspection (drop light)
  —external inspection
  —proper labeling
  —cleanliness.
- Organize containers to minimize the possibility of using the wrong one.
- Ensure that forktruck drivers are aware of the possibility of puncturing containers with the lifting forks.
- Make adequate provisions to prevent runaway rolling drums.
- Place adequate seals on container openings to prevent contamination.
- Inspect incoming containers for leaks.

3.3.4. Emptying of Containers

The emptying of containers can present many hazards, depending on the material being handled, the type of equipment used to empty the containers, and the characteristics of the system into which the material is transferred. Materials hazards are the subject of Section 2.2, and later sections of the present chapter deal with specific equipment. The checklist below should also be reviewed in preparation for handling and emptying containers.

- Provide systems to ensure that the correct material is being unloaded.
- Be aware of the proper method for moving containers to and from skids.
- Be aware of pinching hazards and the possibility of back injury.
- Use protective equipment and protective clothing as required.
- Know the location of safety and alarm equipment and how to use the equipment.
- Equip workers with static-resistant clothing and nonsparking tools when necessary for handling flammable materials.
- Check that the electrical classification of the area is adequate.
- Adequately ground and bond equipment when handling flammable materials.
- Provide for elimination of static discharge potential in flammables.
- Clean the handling equipment to prevent dangerous reactions or contamination.

- Provide systems to prevent release of hazardous materials to the atmosphere.
- Vent containers properly to avoid nonuniform flow or container collapse.
- If necessary, maintain an inert atmosphere while unloading.
- Provide fire-safe or self-closing charging valves if required.
- Check pumps for correct rotation and operation to prevent backflow.
- Determine accuracy of weighing devices and other instruments used to control or monitor transfer operations.
- If thawing is necessary, review the safety of the procedure, and determine if special handling is required after thawing.
- Make sure that Material Safety Data Sheets (Figure 2-1) are on file and available for all materials received in small containers.

3.3.5. Warehousing

The general subject of warehousing is covered in Section 4.3. This discussion refers only to small containers. Warehousing may refer to empty containers, to those received with the plant's consumables, or to purchased containers that have been filled with the plant's product. Some safety considerations refer to the movement and storage of containers, filled or empty; others refer to the characteristics of their contents. The following check list covers both considerations, and also covers movement to and from a warehouse:

- Provide container covers where applicable.
- Specify the maximum safe stacking height.
- Stack containers in a way that permits safe access for fork lift trucks or other vehicles.
- Ensure that skids are adequately constructed and in good condition to assure the stability of containers.
- Train workers properly to avoid damage to containers and spilling of contents during movement.
- Label containers properly and clearly, and inventory them to minimize confusion.
- Use weatherproof labels and ink.
- Segregate materials and containers according to degree of reactivity, flammability, or toxicity as necessary, and confine them to assigned areas in the warehouse.
- Develop procedures for routine inspection of containers.
- Document all inspections to satisfy regulations and to provide accountability and traceability.
- Establish procedures for dealing with leaking containers.
- Ensure that all relevant personnel are trained and tested in these procedures.
- Review the shelf life of the material at storage temperature.

- Determine the temperature at which the material should be stored to avoid uncontrolled reaction.
- Review the methods to determine the activity levels of inhibitors in reactive materials.
- Consider segregating different classes of material:
  —water reactive
  —flammable liquids
  —flammable solids
  —thermally unstable
  —high freezing point
  —pyrophoric
  —toxic
  —regulated materials (e.g., pharmaceuticals, food-grade products).
- Maintain suitable conditions (temperature, humidity, protection from the elements) in the storage area.
- Provide fire protection in the storage area if required.
- Ensure that ventilation is adequate.
- Ensure that the electrical classification of the area is suitable for the type of facility and the materials being stored.
- Provide adequate drainage and spill control.
- Provide physical barriers to prevent the spread of any release into adjoining areas (e.g., walls, dikes).
- Protect critical piping, electrical, and ventilating systems from damage during transport and stacking operations.
- Keep a practical minimum of inventories of dangerous materials.
- Ensure that inventories are turned over periodically.

3.3.6. Loading and Shipping

The activities of loading and shipping containers are primarily the reverse of those already considered in receiving and emptying of containers; therefore, required safety precautions will be much the same.

The points made above relative to the safe handling and transport of containers also apply here. The special concerns in filling the containers relate to the material being handled and are covered by the following checklist:

- Identify the hazards of the material being packaged.
- Properly ground the containers and the filling system(s).
- Determine whether the containers must be inerted because of the reactivity or flammability of the contents.
- Consider the possibility of dust explosion.
- Ensure that ventilation is adequate.
- Set filling levels with due consideration to the effects of thermal expansion and contraction (pressure and vacuum).

- Provide systems to prevent overfilling.
- Ensure that weighing and gauging devices are accurate and certified.
- Specify and apply proper labeling to containers.
- Check product labels, shipping instructions, and shipper's training for adequacy.
- Package the material at a low enough temperature to ensure personnel safety.
- Ensure that the containers have been decontaminated and the filling system adequately cleaned.
- Check that the electrical classification of the area is adequate for the material being handled.
- Seal containers according to regulations and the characteristics of the product.
- Inspect the shipping vehicle to ensure that
  —there are no nails or sharp objects to puncture containers
  —the floor is in good condition and will support the load
  —the roof will prevent rain and snow from entering the storage compartment.
- Make the shipping organization fully aware of the nature of the cargo, its relevant hazards, and applicable emergency procedures.
- Have a trained person available or on call at all times during transit.
- Provide complete information on special product characteristics, such as sensitivity to heat, cold, atmosphere, etc.
- Ensure that the load is properly stowed and secured.

3.3.7. Disposal of Containers

Proper disposal of containers is the responsibility of the user, even if the job is contracted out. The user still must verify that containers have been thoroughly emptied and, where necessary, effectively decontaminated before their final disposal. It is possible to recycle certain containers. Recycling can be done on site or by an outside firm who will reuse or sell the drums.

Thorough cleaning is not possible with bags. The cleanability of flasks, carboys, cans, etc. depends on the type of container. Incineration, scrap recycling, and landfill are still the most frequent methods of final disposal. Container disposal is also discussed in Section 2.7.5.

3.4. PIPING

Chemical processing facilities usually contain many pressurized pipelines. These lines may be at elevated or cryogenic temperatures and may convey flammable, toxic, reactive, or corrosive fluids. Piping systems transporting such fluids must be designed, installed, operated, and maintained properly to reduce the possibility of exposing plant personnel or the community to hazards.

3.4.1. Piping Codes and Specifications

Piping systems are similar to process vessels in that they are required to contain process fluids but unlike process vessels in that the fluids in pipes are moving. The design of piping systems is therefore governed by a different set of codes. Table 3-3 lists some of the piping codes and published literature that should be consulted for safe design.

Piping codes define the design requirements to withstand the mechanical stresses produced by the temperature and pressure of the process fluid. The codes do not address piping material selection and process fluid compatibility other than by recommending that a corrosion allowance be included when choosing the wall thickness of a pipe. Each plant should set its own piping specifications, based on established codes and augmented by specific details, to reduce the chance of failure from localized stress or corrosion. Each piping specification should include as a minimum:

- An identifying number to distinguish it from other piping specifications used in the plant.
- The types of process fluids that can be handled safely under the specification. Usually a single pipe specification can be used for several process fluids. This practice reduces spare parts inventory. the probability of confusion, and the frequency of shortages of critical parts.

TABLE 3-3 List of Piping Codes	
Published jointly by ANSI and ASME:	
B31.1	Power Piping Code
B31.3	Chemical Plant and Petroleum Refinery Piping Code
B31.5	Refrigeration Piping Code
B31.8	Gas Transmission and Distribution Piping Code
B36.10M	Welded and Seamless Wrought Steel Pipe
B36.19M	Stainless Steel Pipe
B16.XX	Various fittings
Published by NFPA:	
NFPA 30	Flammable and Combustible Liquids Code
Published by API:	
API 5L	Specification for Lined Pipe
API 601	Metallic Gaskets for Raised Face Pipe Flanges

- The maximum allowable temperature. For piping in cold service, the minimum temperature also is important.
- The maximum allowable pressure.
- The materials of construction. Different but compatible materials are often used for piping, fittings, and gaskets. All materials used must be resistant to the process fluid. Surface finish or roughness also is specified when product contamination can result from material buildup in crevices.
- The joining technique (flanges, welds, threaded connections, etc.). This may vary with pipe diameter.
- Recommended schedules or wall thicknesses for the various diameters.
- Recommended valve types (gate, globe, ball, plug, diaphragm, etc.). More than one type may be used depending on application (shutoff, throttling, or control) or on pipe size. A list of recommended valve manufacturers may also be included to reduce inventory requirements and the chance of selecting an incorrect valve.
- A list of specialty items (thermowells, expansion joints, rupture discs, sanitary connections, quick disconnects, etc.).

3.4.2. Piping Design Safety

Many factors must be considered when designing a piping system:

- Select the proper piping specification for the fluid to be handled.
- Select the correct pipe diameter to prevent
  —inadequate flow, which could affect quenching or cooling of a reaction
  —choking of gases, which could reduce safety relief
  —flashing of liquids, which could cause cavitation and pump wear or even the possibility of an explosion if flammable liquids are present
  —too low a velocity during slurry transport, which can allow the solids to settle
  —water hammer, which can cause mechanical damage to pipes and equipment
  —slug flow in vapor/liquid systems, which can cause mechanical damage comparable to water hammer
  —erosion due to solids or high liquid velocity
  —plugging or slugging of solids in pneumatic systems because of low fluid velocity, which may lead to dust explosions or personnel exposure to dust irritants
  —static electricity buildup in combustible and explosive fluid systems due to high velocity in nonconductive pipe and fittings
- Check the piping system for undesirable routing or changes in direction that may cause
  —siphoning

—vapor traps in vertical loops, preventing liquid flow or allowing buildup of dangerous vapors

—water hammer

—increased erosion from impingement of solids on walls of pipe and fittings

- Determine the need for check valves or backflow preventers. Make sure that they are correctly oriented and that proper documentation is available to verify this.
- Install safety relief devices on piping and process equipment that may be overpressured.
- Determine whether a rupture disk should be used exclusively in place of a safety relief valve.
- If necessary, install nonfragmenting rupture disks constructed of material resistant to corrosion by the process media beneath relief valves. Fragments from a blown disk can block the valve orifice and prevent reseating. The disk should be positioned upstream of the relief valve to prevent corrosion of valve internals. There should also be some means (e.g., a pressure instrument) of detecting failure of the disk or leakage through a pinhole, with a valve to allow the space between the disk and the safety valve to be vented.
- Size safety relief devices according to the appropriate safety code (see Table 3-4) for the most demanding of these possible upset conditions:

  —closed valves on line or equipment

  —hydraulic expansion of liquid

  —thermal expansion

  —abnormal heat load due to inability to control heat exchange process

  —failure of coolant flow to process

  —failure of reflux

  —failure of pressure control

  —fire in surrounding area

  —runaway reaction

  —infiltration of volatile or reactive foreign material producing a vapor
- Install the relief device as close as possible to the source of pressure in order to minimize upstream pressure drop. Consult the appropriate safety codes on this layout.
- Avoid the use of isolation valves on the inlet or outlet of relieving devices. If a shutoff valve (often referred to as a stop valve) is to be added for maintenance purposes, the installer should consult the applicable codes to determine whether it is permitted. A stop valve must have a port area greater than or equal to the inlet size of the relieving device. Twin relieving devices are often installed with positive isolation provided by three-way valves interlocked so that one relieving device is always in service.

TABLE 3-4
Pressure Relief Codes and Standards

Published by the American Petroleum Institute (API)
Std. 510   Pressure Vessel Inspection Code—Maintenance Inspection, Rating, Repair, and Alteration.
RP 520   Sizing, Selection, and Installation of Pressure-Relieving Devices in Refineries.
          Part I—Sizing and Selection
          Part II—Installation
RP 521   Guide for Pressure-Relieving and Depressurizing Systems.
Std. 526   Flanged Steel Safety Relief Valves.
Std. 2000 Venting Atmospheric and Low-Pressure Storage Tanks (Nonrefrigerated and Refrigerated)
RP 2001   Fire Protection in Refineries.

Published jointly by American National Standards Institute (ANSI) and American Society of Mechanical Engineers (ASME) as the ANSI/ASME codes
B31.1   Power Piping Code.
B31.3   Chemical Plant and Petroleum Refinery Code.

Published by the American Society of Mechanical Engineers (ASME)
Boiler and Pressure Vessel Code, Section VIII, Pressure Vessels.
PTC 25.3 Performance Test Code—Safety and Relief Valves.

Others
NFPA 30
OSHA 1910.106
CGA Standards
These contain guidelines for estimating heat loads due to fire. With regulated materials, OSHA
1910.106 is mandatory.

- Eliminate by-pass lines around relief devices or consult the appropriate safety codes for this type of installation.
- Locate expansion joints in long pipe lines to relieve stresses produced by thermal expansion of the lines.
- Install vents at high points of lines and equipment, as required.
- Install drains at low points of lines and equipment, as required.
- Determine whether vents and drains should be connected to individual piping systems for collection and disposal, rather than being released to the environment.
- For gas, vapor, and steam lines containing unwanted entrainment, install the following at appropriate low points:
  —drip legs
  —traps
  —condensate return systems
  —knockout pots
- For two-phase flow in a vertical riser:
  —Size the line to maintain bubble flow or dispersed flow conditions.
  —Determine whether a dual riser should be installed instead of a single riser (see Figure 3-6).

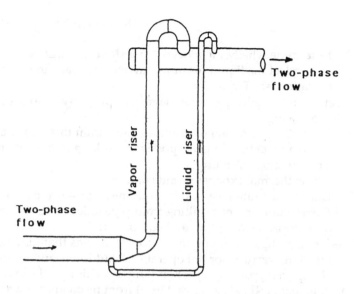

a) Dual riser piping arrangement, used with moderately low vapor velocities to prevent choking, slugging, and resulting water hammer

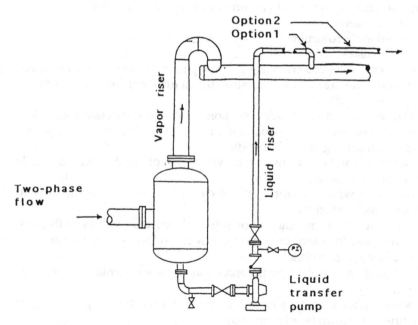

b) Knockout pot arrangement, used when vapor velocity is very low and entrained liquid would choke if it were not disengaged. Option 1 is used when vapor and liquid must enter next process step together. Option 2 maintains separation of phases when that is more desirable.

FIGURE 3-6. Piping Arrangements for Two-Phase Flow.

—Determine whether a knockout pot should be installed to disengage and collect the liquid for pumping in a separate line to the higher elevation (see Figure 3-6).
- Determine whether pipelines have been properly supported to achieve the following:
  —Restrict movement. For example, significant thrusts are produced when a relief valve opens, possibly resulting in movement that may cause damage or injury.
  —Allow thermal expansion and contraction.
  —Dampen vibration caused by equipment or seismic load.
  —Restrict movement resulting from pipe failure and so prevent damage to nearby pipes and equipment.
  —Identify support requirements for conditions that may exist during erection, startup, normal operation, shutdown, and maintenance.
- Make special provisions for handling, installing, and protecting brittle piping (e.g., FRP, glass, or cast iron) from mechanical fracture.
- Determine if the piping contains stress concentration points or unnecessary abrupt changes in direction.
- Insulate the piping system according to specifications for:
  —heat conservation
  —personnel protection
  —anti-sweat
  —process fluid protection (e.g., in pipe passing close to another pipe or through an area at a temperature much different from that of the process fluid)
- Allow enough space for safe operation of manual valves (see Figure 3-7).
- Prevent tampering with isolation and shutoff valves that are not manipulated during normal operation.
  —wrenches and handwheels removed and kept in a location accessible only to authorized personnel
  —key lock with key stored in a secured location, accessible only to authorized personnel
  —lockout feature for motor- or solenoid-operated valves with password or key for valve operation in a secured location, accessible only to authorized personnel
- Position items requiring maintenance in easily accessible locations (see Figure 3-8).
- Install reducers in lines to reduce the potential for air pocketing or sediment buildup (see Figure 3-9).
- Install flame arresters in lines and vents carrying combustible vapors.
- Assess the requirement for double-block-and-bleed valve arrangements—where personnel are occasionally expected to enter vessels—in high pressure steam lines where isolation is needed for maintenance.

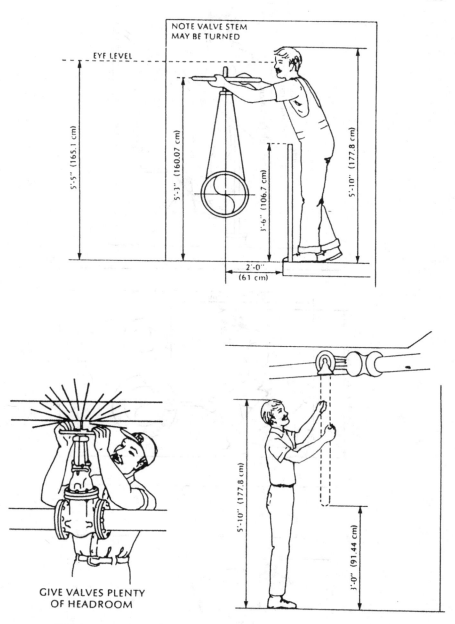

NOTE VALVE STEM
MAY BE TURNED

EYE LEVEL

5'-5" (165.1 cm)

5'-3" (160.07 cm)

3'-6" (106.7 cm)

5'-10" (177.8 cm)

2'-0"
(61 cm)

GIVE VALVES PLENTY
OF HEADROOM

5'-10" (177.8 cm)

3'-0" (91.44 cm)

FIGURE 3-7. Safe Installation of Manual Valves. Locations and Clearances.

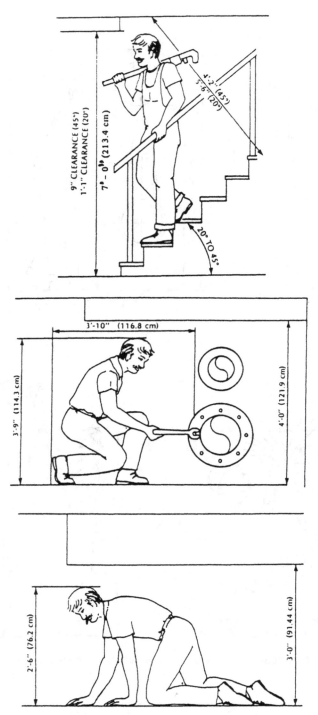

FIGURE 3-8. Maintenance Access Requirements, Walking, Kneeling, and Crawling.

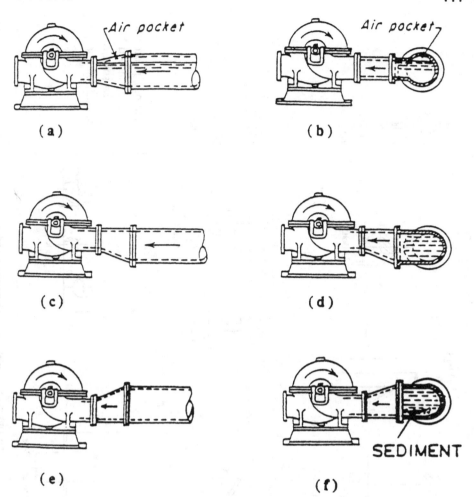

FIGURE 3-9. Installation of Pump Suction Reducers. (a,b) Concentric reducer can allow formation of pocket of undissolved gases when pump is shot off. This produces cavitation at startup. This reducer also can allow sedimentation of hazardous solids in the line during shutdown. (c,d) Installation of eccentric reducer in this orientation eliminates gas pockets. (e,f) This orientation prevents sedimentation.

3.4.3. Piping Installation Safety

Once a piping system is designed, it must be properly installed. Figure 3-10 gives some examples. Some of the items to consider are:

- Install the piping according to drawings and specifications. Any changes must be documented and approved.

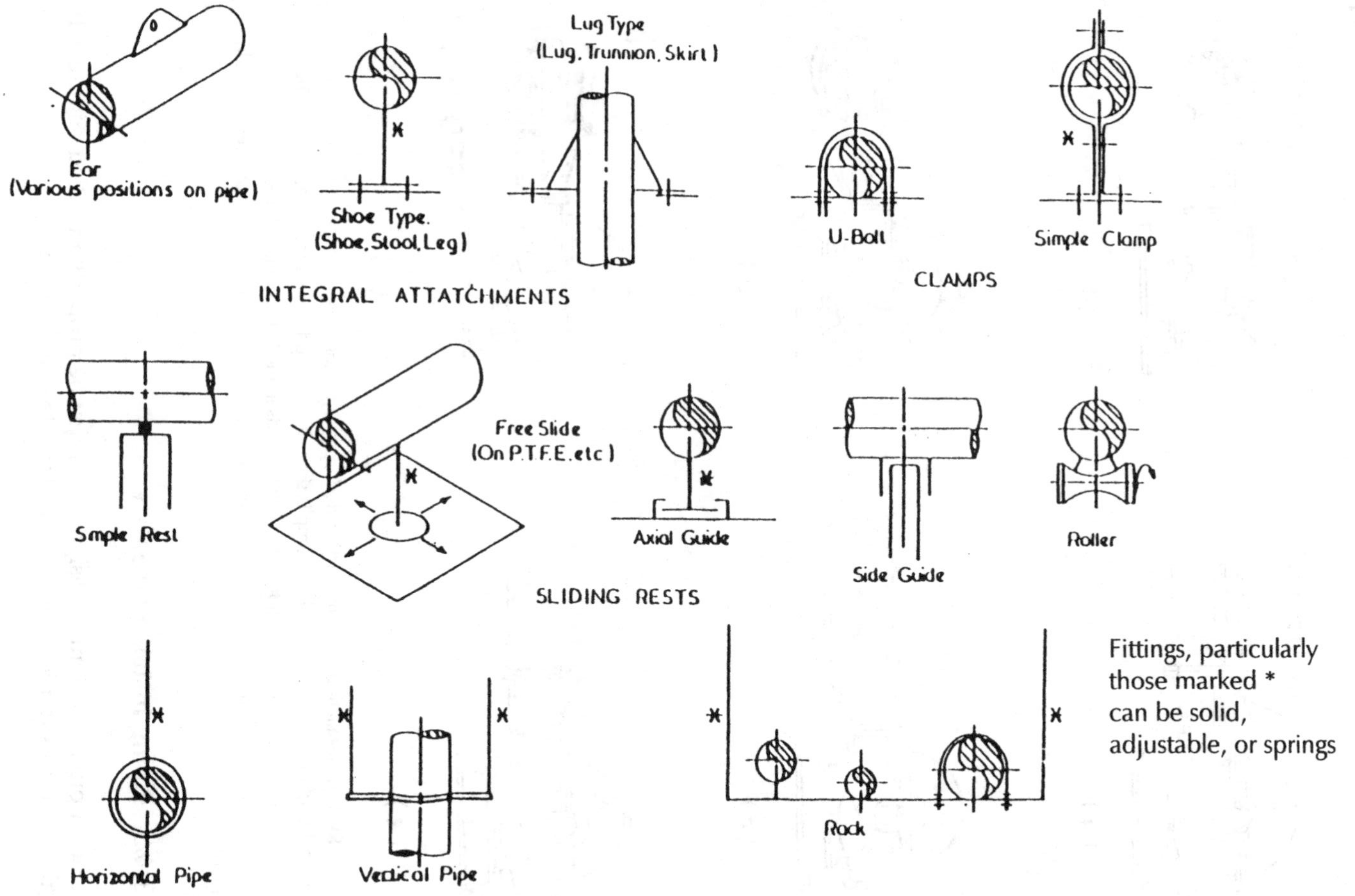

Ear (Various positions on pipe)

Shoe Type (Shoe, Stool, Leg)

Lug Type (Lug, Trunnion, Skirt)

INTEGRAL ATTACHMENTS

U-Bolt

Simple Clamp

CLAMPS

Simple Rest

Free Slide (On P.T.F.E. etc)

Axial Guide

Side Guide

Roller

SLIDING RESTS

Horizontal Pipe

Vertical Pipe

Rack

Fittings, particularly those marked * can be solid, adjustable, or springs

FIGURE 3-10. Examples of Pipe Support Methods.

112

- Install all valves with correct orientation. This is particularly important with check valves.
- Use the correct welding materials.
- Ensure that all the piping has been properly joined and all the fittings have been properly installed.
- Inspect all connections.
- Properly locate all the pipe support hardware. Do not use pipes as supports (see Figure 3-11).
- Leak-test and pressure-test the piping system.
- Install and test all tracing and insulation.
- If painting is required, ensure that it meets specification and plant standards for color coding.
- Place line numbers or lettering describing the process fluid prominently on the pipe, as well as arrows indicating direction of flow.
- Place barriers and shields around piping to prevent damage from collision.
- Place all manual valving in the correct pre-startup position.

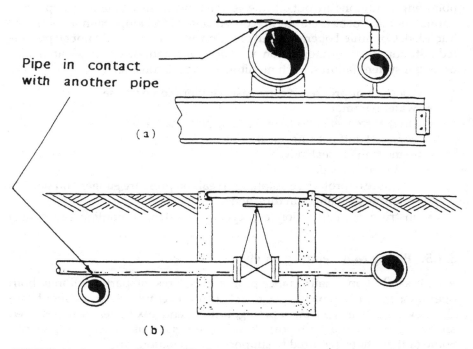

Pipe in contact
with another pipe

(a)

(b)

FIGURE 3-11. Examples of Improper Pipe Support. (a) Pipe using another pipe as support. (b) Underground pipe placed on top of another buried pipe. These arrangements can stress pipes and promote galvanic corrosion.

3.4.4. Piping Operation Safety

Personnel must run a process within established constraints if operation is to remain safe. During normal operation, this requires the following:

- Fluid flows, temperatures, and pressures must be kept within the limits of the operating procedures. Document any variances, intentional or accidental, for investigation. The documentation should describe:
  —deviations from normal flow rate
  —deviations from normal temperature and pressure
  —changes in physical properties or behavior of fluids, such as viscosity, surface tension, density, flashing, condensing, foaming, etc.
  —changes in system behavior, such as excessive vibration or noise, and even unusual reductions in vibration or noise
- Checks on grounding of piping carrying flammable materials
- Systematic monitoring to determine whether harmful substances are being released

Increased production rates may create mechanical stresses on piping and equipment beyond those provided for in design or performance tests. If operating conditions are outside the design range of instrumentation, process parameters cannot be accurately recorded and control loops cannot operate as intended. Continued operation out of the normal range should not be permitted without proper authorization and continual monitoring of operations for potential system failures. Such monitoring should consider

- constant or progressive increases of vibration or noise
- water hammer
- premature wear and failure of components, such as:
  —gasket and packing leaks
  —mechanical seal failure
  —rupture disk bursts
  —failure of protective diaphragm seals on pressure gauges and sensors
  —failure at changes in direction
- instrument malfunction (e.g., cycling, unintended nonlinear response)

3.4.5. Piping Maintenance Safety

A successful piping maintenance program requires the participation of both operating and maintenance personnel. The preventive program should include systematic inspection of piping, fittings, and auxiliaries. Any observed abnormalities should be reported and investigated as quickly as possible. Some of the actions required to support safe operations are:

- Replace all damaged insulation. If a line is traced, repair any damaged portions of the tracing.

- Periodically renew all damaged and aging color coding of piping.
- Periodically replace all damaged or aging line numbers, fluid designations, and flow designations on piping.
- Check for sources of leaks. Make sure that joints are properly seated and that bolted connections are properly tightened.
- Replace or repack leaking valves and flanges. Never reuse old gaskets or packing.
- Keep lines free and clear. Clean or replace strainers when necessary. Check operation of steam traps regularly.
- Document changes in the piping system.

When it is necessary to remove or break into lines, a permit (Section 7.1.4) should be issued. This would establish that all due preparations were made and would require management approval before work began. The specific actions covered would include at least:

- For personnel protection and safety assume the lines to be loaded and under pressure.
- Drain and vent the line and connections to pumps, valves, and tanks.
- Clean or flush line and reduce temperature or pressure as required.
- Lock out all pumps on the line, and check all pumps and sight glasses for zero readings.
- Use appropriate personal protective equipment.
- Place deflectors of suitable materials over flange joints when initially cracking the line.
- Handle removed pipe sections, valves, or components with care, assuming that there are residues or trapped materials.
- Blank off openings in broken lines remaining in place.

Breaking and opening of lines are covered in more detail in Section 4.5.3.

3.5. TRANSFER HOSES

In small plants and especially in those in which certain items of equipment have multiple uses, temporary connections are relatively common. Transfer hoses are often used to make such connections. This practice increases operating flexibility, reduces costs, and often helps to avoid cross-contamination between batches. However, it also presents new hazards in handling of the hoses, potential exposure to dangerous material, possibly wrong connections at hose switch stations, and more frequent leaks. The use of hard piping connections is therefore preferable to the use of transfer hoses. If this is not practicable, there should be, at a minimum, a conscientious program of replacement, inspection, maintenance, and proof of correct connection of hoses.

3.5.1. Materials of Construction

Construction materials can vary greatly, because hoses sometimes are fabricated in exotic materials. The most common materials are stainless steel, rubber, and other synthetics.

Some systems include chemically resistant liners that are physically strengthened by an outer shield or support. The metallic part usually is braided, in order to achieve a combination of strength and flexibility.

When linings are used, their resistance to physical abuse, as well as their chemical properties, must be considered. End fittings suffer more abuse than hose bodies and so deserve special consideration. A coating or a lining suitable for the run of a hose may not be adequate for the wear and tear of continual movement. The end fittings might therefore be made of a resistant metal not practicably or economically available in woven hose construction. An example might be a plastic-coated or plastic-lined hose with Monel end fittings.

Where gaskets or other seals are required with the hose, all the above considerations and other types of sealing devices will apply.

3.5.2. Safety in Design and Installation

There are three fundamental factors to consider in selecting the appropriate hose:

1. temperature limits in service
2. pressure limits
3. corrosion/compatibility with process fluids.

A hose will be rated for certain temperature and pressure conditions when new. Age and service conditions can cause corrosion or loss of mechanical properties and may make it necessary to downrate a hose. The user must have a program for inspection and timely replacement of hoses.

The configuration and flexibility of the hose are important physical considerations. Installation must not produce excessive stress. Each time a hose is used, it should be checked for proper installation, taking the following into consideration:

- The hose must fit the span between connections comfortably.
- The hose must not be distorted after connection.
- Contact with a very hot or very cold surface can impair operation of the hose.
- The hose should not be supported by or in firm contact with a member subject to vibration.
- Stress can be reduced by adding fittings such as elbows or couplings to the connection points.

There are other points to check in addition to the possibility of stressing a hose by inappropriate assembly:

- Check the hose for blockage.
- Double-check visually and by pressure testing all connections where appropriate.
- Check the assembly for leaks.
- Make sure that all fittings and connections necessary for venting, draining, and cleaning are in place.
- Protect the installation externally against corrosion and mechanical damage (falling objects, moving equipment, maintenance activity, etc.).
- Satisfy all requirements for grounding and bonding (see Section 2.2.5).
- Ensure that the installation of the hose is in accord with all company standards.

3.5.3. Safety in Operation

The safety precautions to be taken with hoses are much the same as those recommended for piping systems in general (Section 3.4). Special precautions in installation and more frequent inspection are covered in the sections immediately above. Special actions can also be taken during operation to control hazards. Many of these have to do with controlling the use of hoses to avoid inadvertent mixing of process fluids and the exposure of hoses to the wrong fluids. The following checklist should be consulted to ensure safety in operation:

- Use different hoses for different materials when possible.
- Dedicate hoses to particular services as practicable to avoid problems of incompatibility.
- Label hoses appropriately when identification of contents is desirable.
- Provide segregated storage for used hoses awaiting cleaning and certification.
- Provide separate storage or clear identification (label, color code, special fittings, etc.) for hoses dedicated to a certain group of services.
- Ensure that all personnel wear proper protective clothing.
- Establish procedures for disconnecting hoses, including verification that the hoses have been vented and drained. Ensure that these procedures are monitored and followed.
- Have counteracting systems (e.g., containment or destruction systems) available and ready for use.

3.5.4. Inspection and Maintenance

The frequency with which hoses are added to and removed from the process configuration provides many opportunities for visual inspection. A hose that shows signs of wear, corrosion, or softening is a candidate for replacement. Some of the frequent signs of deterioration are:

- cracking of the outer surface
- corrosion of the braid
- chemical attack on the interior of the hose
- kinks
- damaged or corroded fittings.

If these signs develop quickly or repeatedly in a given service, there may be a problem of compatibility of the hose with the process fluid at normal conditions. In such a case, the user should identify alternative hose materials for trial.

There should also be a preventive maintenance program in which hoses are visually inspected and given physical tests. Chemical deterioration, loss of physical properties, fraying, etc. are possible reasons for replacement. Pressure testing and comparison with new hoses will be useful in determining suitability for service. An alternative or adjunct to the PM program might be the periodic replacement of hoses without regard to their condition.

Transfer hoses are subject to all the hazards that apply to piping (see Section 3.4). Other hazards arise because of their special nature. The following checklist gives some guidance to the user of flexible hoses.

- Install equipment on hose connections to provide a positive means of preventing accidental disconnects due to vibration during normal operation.
- Clean hoses after use to prevent exposure of personnel to hazardous contents and to prevent possible reaction when an incompatible material is introduced.
- Dry hoses as necessary to remove water or other materials that may be reactive with the process fluid (this may best be done immediately before use).
- Check the integrity of the hose (see above for discussion of maintenance and inspection).

3.6. PUMPS

A pump is used to provide motive energy to a liquid in a piping system. Pumps are needed whenever the liquid can not flow at the required rate and in the desired direction because of process conditions or the physical arrangement of the equipment and the piping. A knowledge of the various types of pumps, their characteristics, their limitations, and how they should be selected, installed, operated, and maintained in piping systems has a direct bearing on their safe operation in processing plants.

3.6.1. Pump Types

Figure 3-12 classifies pumps as positive-displacement or kinetic. Positive-displacement pumps move liquid at a relatively constant rate and are used when discharge pressures are high, when the pump must be self-priming, and when liquid must be metered at an accurately prescribed rate. Flow through a reciprocating pump is produced by the movement of a piston or a plunger

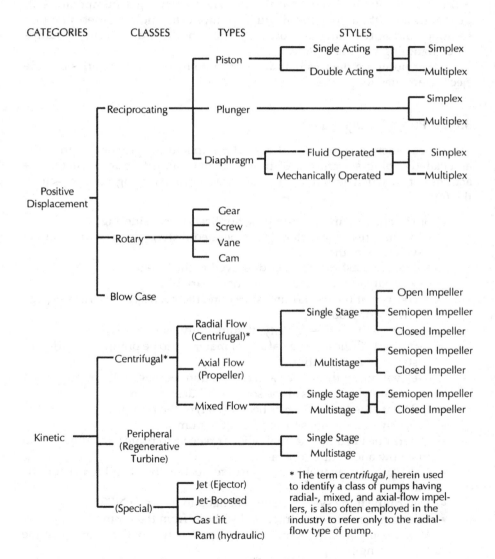

CATEGORIES   CLASSES   TYPES   STYLES

* The term *centrifugal*, herein used to identify a class of pumps having radial-, mixed-, and axial-flow impellers, is also often employed in the industry to refer only to the radial-flow type of pump.

FIGURE 3-12. Classification of Pumps. With permission of the American Institute of Chemical Engineers

through a chamber in the pump casing. The fluid enters and leaves through valves in the chamber. Rotary pumps are most often used with relatively high viscosity liquids that are self-lubricating. Rotary pumps can also produce high discharge pressures and can handle a limited amount of cavitation.

Kinetic pumps transform the energy of the driver into kinetic energy in the liquid. Most pumps found in process plants are centrifugal, because these have the lowest price and the lowest maintenance requirements for most process applications. Only specially designed centrifugal pumps are self-priming, and, although most centrifugal pumps can handle a variety of liquid solutions and slurries, they are not as efficient as positive-displacement pumps when handling viscous liquids.

Other pumping devices that do not fit into these two categories include ejectors and gas-lift pumps.

3.6.2. Pump Design Safety

The procurement of a properly designed pump and the proper design of the associated piping system are of primary importance to pump safety. The actions to take when designing a pumping system and buying a pump include the following:

- Sufficiently define the pumping application, considering:
  —the nature of the material (e.g., aerated water, viscous liquid, slurry)
  —volatility of the liquid
  —possible flashing of gases dissolved in the liquid
  —viscosity of the liquid—constant or variable
  —concentration of solids in a slurry and their physical description (e.g., stringy or globular)
  —material hazards—toxicity, flammability, or explosivity
  —multiple duties, e.g., a variety of materials to be pumped at different stages of an operation
  —elevations of the source and destination relative to the pump
  —operating pressures at the source and destination
- Calculate the total dynamic head (TDH) of the pump.
- Properly size the lines to and from the pump.
- Determine need for external relief or recycle to protect the pump from low flow and shutoff operation.
- Calculate the available net positive suction head (NPSH$_A$) for the pump.
- Determine if the system can be changed to improve NPSH$_A$.
- Eliminate unnecessary traps and siphons from the piping.
- Verify compatibility of materials of construction of the pump with the liquid being pumped.
- Determine if the pump should be self-priming.

- Determine the type of pump to be used.
- Consider the need for a canned motor pump, which is a leakproof sealless pump used with corrosive, volatile, radioactive, and other hazardous liquids. Pumps with magnetic drives have similar advantages.
- Choose pump packing or mechanical seals; establish compatibility of seal fluids with process fluids.
- Select pump driver.
- Once the pump has been selected, ensure that it satisfies all the above factors, including:
  —TDH provided by the pump greater than the TDH required by the application
  —net positive suction head required by the pump (NPSH$_R$) less than the calculated NPSH$_A$.

3.6.3. Pump Installation Safety

Some of the requirements for safe installation of a pump are the following:

- proper equipment and number of qualified personnel available to move the pump into position
- pump pad, including bolting, installed correctly
- location and orientation of pump match those shown on piping drawings
- pump mounted correctly and aligned with driver
- packing/mechanical seal and piping to and from the pump checked for leakage
- auxiliary systems for external cooling or lubrication of packing or mechanical seal properly installed and checked out, and ready for operation
- all guards for rotating parts properly installed
- electrical and motor enclosures suitable for the application
- driver properly hooked up (e.g., no reversed wiring)
- pump, driver, and lifting lugs located to assist access and maintenance
- pump located away from ignition sources
- location and installation to allow isolation from source of pumped liquid in case of fire

3.6.4. Pump Operation Safety

Operation of equipment away from its design point can produce stresses that lead to premature failure. Pumps are a good example. Many are able to operate well above design flow if the discharge head is reduced. This can happen when a pump is transferred to a new duty. The result often is surging and vibration.

Use of a pump at a flow rate below design may seem innocuous. However, the resulting internal recirculation induces hydraulic vibrations. These have many harmful effects, including rapid wear of components and the possibility of release of fluid through a failed seal or bearing [9]. Other signs of improper selection or operation are

- pump discharge pressure too high
- current draw on the pump motor too high
- little or no liquid flow through the pump
- pump operation noisier than expected
- pump packing or mechanical seals leak and require more maintenance than expected
- pump internals, such as impellers, plug frequently or wear away prematurely
- pump bearings wear rapidly
- pump shaft frets

Operation of a pump with the discharge shut off forces all the frictional heat into a small volume of liquid and can raise its temperature to dangerous levels. Note that provision of a minimum-flow bypass around the pump may not prevent this situation if the fluid is returned directly to the suction of the pump.

3.6.5. Pump Maintenance Safety

Safe operation of pumps requires an understanding of the problems that can occur, their causes and possible corrective actions. The maintenance program can be built around the list of operating problems and corrective actions in Table 3-5.

3.7. FANS AND COMPRESSORS

Fans and compressors are gas movers, analogous to pumps as liquid movers. The term "gas mover" will be used when discussing the characteristics of fans and compressors collectively. The geometry of gas movers is somewhat similar to that of pumps; however, operating parameters and safety considerations are quite different. Fans and compressors must handle gases that are compressible and extremely sensitive to temperature and pressure changes, while pumps handle liquids that are relatively insensitive to pressure and temperature changes and can be considered incompressible.

3.7.1. Classification of Gas Movers

Figure 3-13 shows the classification of gas movers into two broad categories, based primarily on maximum discharge pressures. The pressure ratio of a fan (absolute discharge pressure to absolute suction pressure) is less than 1.2. The

TABLE 3-5
Pump Operation: Problems, Causes, and Cures

Consequential Safety Hazard	Cause	Solution
NO LIQUID FLOW		
• Recipe error. • Loss of process control. • Runaway reaction. • Relief valve opening and discharge of hazardous material. • Loss of cooling and consequent overheating.	Suction or discharge valves closed.	Open valves.
	Pump is not primed.	If priming system has not been installed, install temporary hookup and design a permanent priming system if problem persists.
	Pump suction is not immersed in liquid.	Solution similar to pump priming situation. Flood suction with liquid using one of pump priming techniques.
	Available NPSH is too low.	Any of the following: • increase suction line diameter • decrease suction line length • increase height of liquid in source vessel • increase pressure in source vessel • remove entrained gases in liquid • cool liquid to lower its vapor pressure • replace with pump requiring a lower NPSH
	Strainer in suction line clogged.	Clean strainer. If problem persists, determine the source of debris in strainer and eliminate it or install a strainer with larger mesh (or install pump that can tolerate the solids load.
	Bypass valve is open.	Close valve.
	Relief valve jammed open.	Repair or replace relief valve.
	Relief valve is set to open at a discharge pressure lower than required for safe pump operation.	Adjust relief valve or replace with a valve having a higher set pressure.
	Pump rotates in wrong direction.	Rewire pump motor correctly.

TABLE 3-5 (Continued)
Pump Operation: Problems, Causes, and Cures

Consequential Safety Hazard	Cause	Solution
PUMP FLOW LOWER THAN RATED CAPACITY		
Same as No Liquid Flow symptom.	Suction or discharge line partially closed.	Verify that valves are fully open.
	Air leakage in suction line or through pump seal.	Check for leaks and repair.
	Available NPSH is too low.	Same solutions as for no flow symptom described on page 123.
	Strainer in suction line partially clogged.	Same solutions as for no flow symptom described on page 123.
	Bypass valve is open.	Close valve.
	Relief valve jammed open.	Repair or replace relief valve.
	Relief valve is set to open at a discharge pressure lower than required for safe pump operation.	Adjust relief valve or replace with a valve having a higher set pressure.
	Pump speed too low.	Options: • Install a gear reducer or variable speed drive to increase pump speed. • Install motor with higher rpm.
	Wear on pump causing increased clearance and fluid slip.	Options: • Reduce pump speed to reduce wear; however, this will reduce pump head. • Install pump constructed of material with greater wear resistance. • Reduce or eliminate material in fluid causing the excessive wear.
	Liquid viscosity is higher than specified for pump.	Options: • Reduce fluid viscosity by heating or diluting. • Install a larger impeller if possible. • Replace centrifugal with rotary pump.

TABLE 3-5 (Continued)
Pump Operation: Problems, Causes, and Cures

Consequential Safety Hazard	Cause	Solution
PUMP LOSES PRIME DURING OPERATION		
Same as No Liquid Flow symptom on page 123.	Liquid level at source falls below suction line.	Install controls or operating procedures to prevent the drop of liquid level.
	Liquid vaporizes (flashes) in suction line.	Follow procedures recommended for improving NPSH.
	Air leakage in suction line or through pump seal.	Check for leaks and repair.
	Dissolved air or gases flash out of liquid.	Follow procedures for improving NPSH.
NOISY PUMP OPERATION		
• Cyclic release from relief system. • Incipient breakdown of pump or development of leak.	Cavitation.	Same as above.
	Foreign material in the pump.	Options: • Remove material from feed if pump cannot discharge it. • Install suction strainer. • Check suction line and source to determine whether additional debris exists and whether it can be removed from the system.
	Misalignment.	Check and correct pump/driver system alignment.
	Bent rotor shaft on rotary pump.	Repair or replace damaged shaft.
	Relief valve chattering.	• Relief valve set point too close to operating point of pump. • Reset relief valve or replace with another valve with higher set point.

TABLE 3-5 (Continued)
Pump Operation: Problems, Causes, and Cures

Consequential Safety Hazard	Cause	Solution
RAPID PUMP WEAR		
Premature failure and release of fluid.	Corrosion	Options: • Add corrosion inhibitor to fluid. • Use less corrosive liquid in pump. • Replace pump with material more resistant to pumped fluid.
	Cavitation: pump running dry.	Follow procedures recommended above for insufficient NPSH problem.
	Grit or abrasive material in liquid.	Options: • Remove abrasive material from fluid. • Reduce pump speed to reduce wear (this will also reduce pump head). • Install pump constructed of material with greater abrasion resistance.
	Stresses imposed on pump by pipe loads.	Examine pipe design to determine stress concentration point. Redesign piping to eliminate loads.
POWER CONSUMPTION TOO HIGH		
• Reduction of flow, with results as noted above. • Overheating of liquid, resulting in more rapid corrosion or boiling in the pump. The latter can cause total loss of flow. • Overheating of motor and power cables.	Obstruction in pump or discharge line.	Eliminate obstruction.
	Valve in discharge line partially or completely closed.	Open closed valve(s).
	Discharge line is too small.	Replace with larger diameter pipe.
	Liquid viscosity higher than specified.	Options: • Reduce fluid viscosity by heating or diluting. • Replace centrifugal with rotary pump.

TABLE 3-5 (Continued)
Pump Operation: Problems, Causes, and Cures

Consequential Safety Hazard	Cause	Solution
POWER CONSUMPTION TOO HIGH (Continued)		
	Misalignment.	Check and correct pump/driver system alignment.
	Shaft packing too tight.	Reduce packing tightness. If leak rate from packing unacceptable, replace packing with mechanical seal.
	Pump speed too high.	Options: • Install gear reducer or variable speed drive. • Install motor with lower rpm (also reduces discharge head of pump).

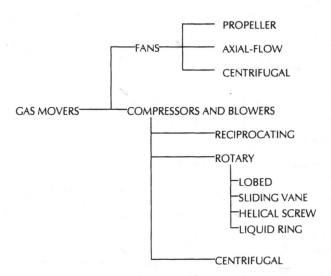

FIGURE 3-13. Classification of Gas Movers.

pressure ratio of a compressor is 2.0 or higher. A blower is a low-pressure compressor with a ratio between about 1.5 and 3.0.

Fans are typically used in process plants to move large volumes of gas with very little increase in pressure. The three basic fan types are propeller, centrifugal, and axial-flow. The latter two are generally used in process applications, while the propeller type is used primarily for air circulation.

Fans have operating curves similar to those of centrifugal pumps but have operating limitations comparable to those of centrifugal compressors. Safety starts with matching the fan to the application. Table 3-6 lists some typical applications. The Air Movement and Control Association (AMCA) provides standards for installing, operating, maintaining, and troubleshooting fans.

TABLE 3-6
Typical Industrial Applications for Fans[a]

Application	Axial-flow Fans		Radial	Centrifugal Fans		Airfoil
	Tube-Axial	Vane-Axial		Forward-Curved	Backward-Inclined	
Conveying systems			✔		✔	
Supplying air to burners or furnaces	✔	✔	✔	✔	✔	✔
Boosting gas pressure			✔		✔	✔
Ventilating process plants	✔	✔			✔	✔
Boilers, forced-draft		✔			✔	✔
Boilers, induced-draft			✔	✔		
Kiln exhaust			✔	✔		
Kiln supply		✔			✔	✔
Cooling towers	✔					
Dust collectors and electrostatic precipitators			✔	✔		
Process drying	✔	✔	✔		✔	✔
Reactor off-gases or stack emissions		✔	✔			

[a]Adapted from Pollack [10]

The American Petroleum Institute (API) publishes standards for specifying, operating, and maintaining compressors. While these standards were developed for refinery service, most apply to other types of plant, and they have become recognized standards in other industries as well. They include:

API 611   General Purpose Steam Turbines for Refinery Service
API 612   Special Purpose Steam Turbines for Refinery Service
API 613   High-Speed Special Purpose Gear Units for Refinery Service
API 614   Lube and Seal Oil Services
API 616   Combustion Gas Turbines for General Refinery Service
API 617   Centrifugal Compressors for General Refinery Service
API 670   Noncontacting Vibrating and Axial Position Monitoring Systems

3.7.2. Gas Mover Operating Parameters

3.7.2.1. Basic Parameters
The basic parameters to consider when purchasing, operating, or maintaining any gas mover are:

- *Volumetric flow rate*—volume of gas flowing at the actual inlet temperature and pressure.
- *Gas composition*—constituents of the gas:
    —pure air or individual components; molecular weights; proportions
    —moisture
    —other condensibles; reactivity; corrosivity; other hazards
    —dust in the gas
    —quantity; friable or abrasive; tendency to agglomerate; hazards
- *Discharge static pressure*—static pressure at the discharge of the gas mover.
- *Equipment orientation and layout*—used by vendors to choose the best gas mover that fits the space limitations.
- *Efficiency*—fraction of the energy applied to the driver that is converted into useful work.
- *Noise*—expected noise level is provided by the equipment vendor in the specification. Noise level may influence selection of the location of the gas mover.
- *Compressor/fan speed*—rate of rotation of the gas mover. This is determined by the equipment vendor to produce the required flow rate and discharge pressure. Rotational speed will in turn affect the amount of noise generated. If variable speed is desired, the vendor should supply performance data for the range of operating speeds.
- *Linear discharge velocity*—linear velocity of the discharged gas will affect noise and frictional losses.

- *Brake horsepower*—power required to produce the required results. This information is provided by the equipment vendor.
- *Vibration limits*—should be set by the vendor. In the absence of such data, a standard list [10] may be consulted. Vibration is often an important factor in the diagnosis of operating problems.

3.7.2.2. Additional Parameters

Gas movers have three other operating parameters that affect their safe operation: heat of compression, surge, and stonewall.

Heat of Compression

Fans do not produce significant heats of compression. Compressors, however, with their higher pressure ratios, can produce large amounts of heat, with higher exit temperatures and marked changes in density. Higher temperature means higher internal energy and therefore more power consumption. If additional power is unavailable or the increased temperature unacceptable, interstage coolers can be used to reduce exit temperature. It is important to establish the highest internal temperature that can be tolerated in a compressor. This temperature can be the service limit of one of the materials of construction (e.g., in a gasket) or the threshold for reaction of some component with the gas.

Surge

Surge is a phenomenon that occurs at conditions of low flow in centrifugal gas movers. All manufacturers include in their equipment specification a minimum allowable flow. At low flow, the dynamics of the piping system and the operating curve of the gas mover can be out of balance, causing pressure pulses and equipment vibrations, known as "surge." The vibration is translated to the thrust system and will ultimately destroy the equipment if surging is not stopped.

Surge can be eliminated by installing a recycle line. A control valve (or louvers) would be designed to open whenever the downstream flow drops below the specified minimum. With recycle to the suction, the flow through the gas mover itself is increased, preventing surge. Anti-surge systems vary from simple open/shut systems for small simple gas flows to highly complex controls for large and extremely variable gas streams. Basic texts [11] describe many anti-surge control techniques, and machine vendors should provide their specific requirements.

Stall

Stall, or incipient surge, produces a noticeable drop in flowrate, while surge does not. This drop occurs when the gas stream actually separates from the impeller. Centrifugal fans and axial-flow fans are susceptible to stall at flow rates just below that which produces the peak discharge pressure. Stall therefore often occurs during startup or when flow is increased too quickly. In multistage compressors, one stage may experience stall while other stages do not. The user should obtain information from the vendor concerning the gas mover's susceptibility to stall and the recommended procedures for overcoming it.

Stonewall
"Stonewall" is a term used to describe choking at the discharge of a gas mover, as evidenced by a rapid drop in discharge pressure at a volumetric flow rate close to the maximum that the gas mover can produce. Stonewall usually occurs when the density of the inlet gas drops. This leads to higher velocities and choking at the discharge. The usual causes of the drop in density are lower inlet pressure, higher inlet temperature, and lower molecular weight of the gas. If these changes cannot be avoided, the gas mover can not meet the process flow demands, and a larger unit is necessary.

3.7.3. Gas Mover Safety Precautions

Because of the similarity of different types of gas movers, many of the same safety precautions apply to all. As compared with fans, blowers and compressors produce higher pressure, generate more heat through compression, and are subject to more damage from vibration. Compressors therefore require more safeguards. Common sources of hazard are surge, stonewall, inadequate alignment or lubrication, and incompatible or leaking seals.

3.7.3.1. Design Safety
Proper design of a machine and its associated piping (or ductwork) is of primary importance to safety. Consider the following factors when designing a system:

- Sufficient definition for the application. The designer must consider
  —nature of the gas (e.g., molecular weight; single gas or a mixture)
  —condensibles in the gas; condensation temperature and pressure; toxic or corrosive condensibles
  —dust in the gas; quantity; corrosive or abrasive
  —whether the gas is toxic, flammable or explosive
  —variance of inlet gas temperature, suction or discharge pressure, or composition
  —potential surge or stonewall conditions
- Proper sizing of connecting lines or ductwork to prevent choke points in the suction or discharge which will force a compressor into surge.
- Requirements for flow straighteners at a fan suction or discharge, or in any bends in the ductwork, to prevent eddy current pressure losses.
- Adiabatic head and required discharge pressure calculations.

- Requirement for anti-surge protection.
- Compatibility of the equipment and materials of construction with the gas.
- Type of packing or mechanical seal.
- Provision of drain connections for moisture-containing gases.
- Spark-proof construction for flammable gas. Standard 99-0401-66 of the AMCA defines three types of spark-resistant construction
  Type A: all parts in contact with the gas nonferrous.
  Type B: nonferrous wheel and nonferrous ring about the opening through which the shaft passes.
  Type C: shift of wheel or shaft will not cause two ferrous parts to rub or strike each other.
- Requirement for special piping connections on the housing to quench or smother an ignited gas.
- Requirement for special connections to ground the equipment or remove static electricity.
- Type of driver required. If variable speed is necessary, the transmission system must properly match the driver to the equipment.
- Suitable warnings on lockout switches.
- Indication, alarm, or lockout on an access door to prevent entry while the fan is operating or coasting to a stop.
- Maintenance access to housing of fans handling dust; access panels for cleaning wheels or replacing wear plates and housing liners.
- Maintenance access to blower and any points likely to plug in pneumatic conveying systems; wear plates may be required with abrasive materials.
- Need for interstage cooling or aftercooling.
- Need for lube oil system or nonlubricated components (e.g., PTFE piston rings, packing).
- For lube and seal oil systems, provision of interlocks to prevent startup of the machine unless the oil system is operating; backup systems.
- Alarms (or shutdowns) in case of high bearing temperature and low oil flow.
- Location of large machines (ground floor best).
- Noise dampening by silencers or acoustic housing.
- Vibration sensors and alarms or shutdowns for large machines.

3.7.3.2. Installation Safety
Some of the factors that must be considered when installing the selected machine have already been discussed in the introduction to this chapter. Additional points are given below:

- Check connections and seals for leaks.
- Correctly align machine with the driver and its transmission.
- Ensure that motor enclosures and other electrical equipment are compatible with the electrical hazard rating of the area.

- Ensure that the driver is hooked up properly (e.g., motor checked for correct wiring and direction of rotation).
- Provide guards on shafts and couplings to prevent injury or contact with foreign objects.
- Satisfy any need for permanent noise abatement.
- Provide insulation on hot compressed gas discharge piping.
- Protect personnel by placing a grating or louvers across the discharge if possibility exists for contact with blades.
- Locate equipment properly. According to OSHA, fans located less than seven feet above a floor require some sort of personnel protection. Safety guards should be mounted around accessible fans. This includes exposed fan inlets and outlets, coupling guards, shaft and bearing guards, and belt drive guards.
- Consider use of open compressor buildings to avoid accumulation of released vapors.

3.7.3.3. Operating Safety
AMCA Publication 202-88 provides troubleshooting guidance. The following should be monitored:

- discharge pressure
- current and voltage draw on the motor
- gas flow through the machine
- surging conditions
- obstructions or restrictions in the connecting lines
- vibration or noise due to fan surging or misalignment
- lubrication
- seals—leakage or excessive maintenance
- internals—frequent plugs or excessive wear (e.g., because of dust loading)
- bearing wear or shaft fret
- contamination of oil
- alignment of drive

A high motor temperature can indicate obstruction of cooling air flow or, if accompanied by high amperage, an increase in the gas load. High bearing temperatures can indicate misalignment or improper lubrication.

3.7.3.4. Maintenance Safety
A good maintenance program is necessary for continued safe operation of any gas mover [12]. The preventive maintenance program should include
- periodic addition and change-out of bearing oil
- periodic clean-out of fans
- examination of couplings for misalignment, cracks, deformities, etc.
- replacement of damaged or missing insulation
- replacement of damaged or leaking sealing devices

Items that may affect compressor performance and safety should be checked regularly:

- efficiency losses
  —valve losses (most significant)
  —ring slippage
  —packing leakage
- cylinder maintenance and wear
- malfunctions
  —exceeding allowable rod load
  —accelerated wear and scuffing
    –piston to liner
    –piston rings
    –piston rod and packing
    –valve breakage
    –knocks, noises, and vibrations
- surge tanks or accumulators for pressure pulse reduction

When a machine is shut down for maintenance, the following should be considered:

- purging all toxic, corrosive, flammable, or otherwise dangerous material from the system
- deenergizing the driver
- securing and locking the fan wheel to prevent "windmilling," which can cause physical damage and injury

After maintenance and before restart, the following should be checked:

- no foreign material in the machine
- proper reinstallation and secure mounting of components
- freedom and direction of spin of impeller
- security of electrical grounding
- alignment
- tension of drive belts
- tightness of connections (mechanical and process)
- personnel guards

3.8. DRIVERS

3.8.1. Motors

Most drives found in process plants will be three-phase squirrel-cage induction motors; a small minority of nonelectric drives are used. Among the electric drives are relatively few synchronous motors; these are usually chosen for large loads or for applications where a low motor speed (e.g., less than 600 rpm) is desired.

The induction motor works by induced currents in the rotor, which produce a force causing the rotor to move in the same direction as the impressed field. In a real system, the rotor is unable to keep up with the field; there will be some "slip" which increases with the amount of torque developed. A two-pole motor driving a centrifugal pump, for example, may turn at 3500, rather than 3600, rpm.

3.8.1.1. Motor Rating and Area Classification

Motors are familiar to all operations personnel, and most will have had a great deal of experience with motors. Each motor application is different, however, and requires a specific review. This is especially important when making changes in operation or when installing stock motors.

The major organizations that publish relevant standards are discussed below. Conscientious application of these thorough and well developed standards will avoid most of the hazards associated with electrical supply and motor operation.

The Institute of Electrical and Electronics Engineers (IEEE) defines rating methods, classifications, test codes, etc. ANSI standards cover much of the same material and set standard dimensions, material specifications, etc. The National Electrical Manufacturers Association (NEMA) also issues standards that are universally recognized in the industry. These are directed more towards specific products. NEMA Standard MG1-10.36 covers motor performance, dimensions, tolerances, frame size, enclosure, and torque classification. Figure 3-14 will help the reader to understand the importance of torque classification. The maximum torque encountered (pullout) may be much higher than the normal operating value, and this fact must be recognized in design in order to prevent mechanical failure. The specific example is that of a centrifugal pump driven by a squirrel-cage induction motor. Each has its own speed–torque characteristic curve. The intersection of the two curves is the normal full-load operating point. It is important to follow the motor's curve down to zero speed to identify the maximum torque.

Table 3-7 shows the NEMA classification of motors and their typical applications. These can have greatly different torque characteristics, as illustrated by Figure 3-15.

Among the ANSI standards is the National Electrical Code (NEC) which is directed at the elimination or control of electrical hazards. It defines the classification of areas according to the degree of hazard possibly found in the atmosphere and prescribes wiring methods, grounding, control, etc. These factors are important in motor selection. The NEC is a product of the National Fire Protection Association (NFPA).

Table 3-8 outlines the NEC definitions of hazardous areas. There are three aspects to the definition. The "Class" defines the nature of the hazard. Class I,

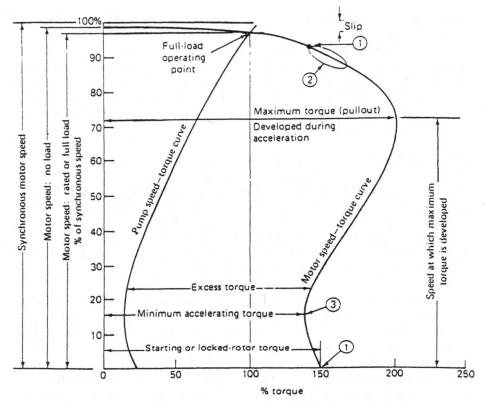

FIGURE 3-14. Performance Characteristics of Motors. For typical squirrel-cage induction motors. Points ① show the speed–torque limits during initial acceleration. Area ② is the maximum torque range recommended for short-term overloads. Point ③ is the reverse-curvature point on the speed–torque curve and is the point of minimum accelerating torque.

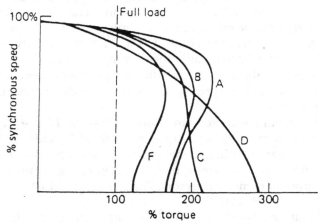

FIGURE 3-15. Torque Characteristics of Motors. Letters correspond to NEMA classes in Table 3-7

TABLE 3-7
Typical Applications of Motors[a]

NEMA Class	Type	Typical Application
A	Normal	Shredder, rotary vacuum pump.
B	General purpose	Pump, blower, centrifugal or rotary compressor, low-viscosity liquid mixer, fan.
C	Double-cage, high-torque	Ball mill, loaded reciprocating compressor, large conveyor, hammer mill, high-viscosity liquid mixer, kneader, rotary kiln.
D	High-resistance rotor	Centrifuge, crusher with flywheel, drum dryer, oil-field pump with flywheel, rotary kiln.
F	Double-cage, low-torque	Uneconomical for most industrial loads because of their high starting-torque to breakdown-torque ratio.

[a]Extracted from Charles C. Libby, Motor Selection and Application, McGraw-Hill Book Company, New York, 1960, Table 5-5, pp. 208–217.

for example, is concerned only with gases and vapors. The "Division" accounts for the prevalence of the hazard. Only if the hazard is present under normal conditions of operation is the application rated in Division 1. Finally, the "Group" designation allows a more specific definition of the hazard within each class.

3.8.1.2. Enclosures

Selection of the proper enclosure is an important step in choosing a motor. Protection of the windings from fumes, splashing fluids, and airborne solids is an important safety factor. The supplier of a piece of equipment who purchases the drive from a motor manufacturer is not always aware of the environment in which the combination will be used. The plant operator who is purchasing driven equipment must consider its surroundings and check the adequacy of the motor, its enclosure, and the transmission (see Section 3.8.3).

Table 3-9 lists NEMA's defined types of motor enclosure and describes their main characteristics, and Table 3-10 lists the NEMA enclosures for other electrical equipment. Since sparks from electrical sources are the most common source of ignition of flammable material (Table 2-2), it is important to select the proper NEMA enclosures for motors and electrical contacts.

Particularly in an indoor application, the noise level produced by a motor should also be considered. This is most likely to be a problem at high rotational speed (3600 rpm).

TABLE 3-8
NEC Hazardous Area Classification
Outline of Articles 500, 501, 502, and 503 of the 1984 National Electrical Code (NEC) Standards*

		HAZARDOUS ATMOSPHERE GROUPING	
Class	Atmosphere type	Hazard group	Typical materials in group
I	Gases and vapors	A[a]	Acetylene
		B[a]	Hydrogen, arsine, ethylene oxide, propylene oxide, etc.
		C[b]	Carbon monoxide, ethylene, cyclopropane, ethyl ether, hydrogen cyanide, etc.
		D[a]	Acetic acid, ammonia, benzene, ethane, alcohol, gasoline, methane, etc.
II	Combustible dust	E	Combustible dusts having a resistivity less than 10^5 ohm cm
		G	Combustible dusts having a resistivity more than 10^5 ohm cm
III	Flammable fibers & flyings	. . .	Carbonized or excessively dry organic materials, ignitable fibers.

	DEFINITION OF CLASSES AND DIVISIONS	
Class	Division 1	Division 2
I	Hazardous under normal operations and normal maintenance.	Hazardous only during abnormal or unusual conditions (e.g., an accident); applies to gases and vapors in closed or well-ventilated systems.
II	Same as Class I or where dust may cause abnormal operation such as a mechanical failure or a short circuit.	Hazardous only when dust is in sufficient quantities to accumulate, causing either a heat buildup by its insulating qualities or arcing of contacts or causing a fire around the equipment.
III	Same as Class I with the qualification that the hazard is caused by combustible fibers or flyings produced during normal handling and manufacturing.	Combustible fibers and flyings not expected to be suspended in hazardous quantities in the air such as in storage and handling (not manufacturing) facilities.

* National Electrical Code, 1984 ed., NFPA 70-1984, published by the National Fire Protection Association, Batterymarch Park, Quincy, MA, 02269.
[a]Assumed lowest ignition temperature of materials is 280°C (536°F).
[b]Assumed lowest ignition temperature of materials is 180°C (356°F).

TABLE 3-9
NEMA Motor Enclosures (adapted from Reference 2, Table 17.6)

Type	Description
I. Open enclosures	Ventilated by openings for air cooling of winding.
A. Dripproof	Ventilation openings designed to prevent most liquids or solid particles falling on enclosure at an angle <15° from the vertical from entering enclosure.
B. Splashproof	Ventilation openings designed to prevent most liquids or solid particles falling on enclosure at an angle <100° from the vertical from entering enclosure.
II. Totally enclosed motors	Totally enclosed to prevent exchange with the atmosphere; however, enclosure is not airtight.
A. Totally enclosed nonventilated (TENV)	No external means of cooling are included with the enclosure.
B. Totally enclosed fan-cooled (TEFC)	The motor is cooled by a fan integrally mounted to the enclosure blowing cooling air over the enclosure.
C. Explosion-proof	The motor enclosure is designed to withstand any internal explosion caused by ignition of flammable gases or vapors from the surrounding atmosphere.
D. Dust-explosion-proof	Motor enclosures suitable for Class II locations, which withstand explosions or ignition of dust on or around the enclosure.
E. Waterproof	The enclosure prevents the entrance of water in the form of a directed stream, such as a hose, with the exception of leakage through the shaft. It may also include automatic draining facilities.

3.8.1.3. Installation and Operation

In the installation of a motor, proper support is essential. Selection of the support is simplified by the use of the various standards listed above. The standards themselves and the suppliers' recommendations can serve as guides. Guards are required where personnel may come in contact with a rotating member. The installer should check specifically for correct rotation and for good alignment. The latter involves whatever series of devices is used to transmit the motor's action to the load (see Section 3.8.3). The preventive maintenance program should follow all the suppliers' recommendations and should include inspection and testing, lubrication and impregnation, cleaning, and protection and safety measures.

Finally, each motor should be checked for the nameplate data required by NEMA. These include

TABLE 3-10
NEMA Classification of Electrical Enclosures
(adapted from Reference 2, Table 17.4)

Type Number	Description	Application
1	General-purpose	Suitable for normal indoor atmospheres; used to prevent accidental contact with the enclosed equipment.
2	Driptight	Same protection as above except that the enclosure must also protect equipment from dripping water and condensate.
3	Weatherproof (weather-resistant)	Suitable for outdoor use; protects equipment from specified weather hazards.
3R	Raintight	Protection from heavy rain (not sleet) exposure.
4	Watertight	Protection for equipment subject to direct streams of water used during cleaning operations.
5	Dusttight	Protection from nonexplosive or combustible dust.
6	Submersible	Permits successful operation of equipment while completely submerged in water for specified conditions of time and pressure.
7*	Hazard Groups A, B, C, or D, Class I (air break)	Suitable for use in NEC Class I hazardous locations. Contact operations and circuit interrupts occur in air environments.
8*	Hazard Groups A, B, C, or D, Class I (oil-immersed)	Same as NEMA 7 enclosure above except that the equipment within the enclosure is immersed in oil.
9*	Hazard Group E or G, Class II	Suitable for NEC Class II dusty atmospheres.
10	Bureau of Mines	Explosion-proof enclosure suitable for use in gassy coal mines.
11	Acid- and fume-resistant	Equipment within this enclosure is immersed in oil. Enclosure provides protection from acid and other corrosive fumes.
12	Industrial use	Enclosure provides protection from dust, fibers, flyings, oil seepage.
13	Dustproof	Enclosure is designed to prevent entry of dust, which may cause contact or interference with the equipment.

*NEC Standards should be consulted for the proper application of these enclosures.

1. Manufacturer's type and frame design.
2. Power output.
3. Time rating.
4. Maximum ambient temperature. (Temperature rise may be substituted.)
5. Insulation system designation.
6. Speed at rated load in rpm.
7. Frequency. If two frequencies are shown, the data in items 2, 3, 4, 6, 9, 10, and 13 must be given for each frequency.
8. Number of phases.
9. Rated load (amps).
10. Voltage
11. Code letter indicating locked rotor kVA/hp
12. Design letter indicating torque and slip characteristics
13. Service factor (if other than 1.0)
14. Thermal protection

3.8.2. Steam Turbines

A steam turbine is a rotary equipment driver that uses steam as its source of energy. Safety considerations relate to the intended use as well as to the selection, installation, and operation of the turbine. Steam turbines are used either as emergency backup to critical motor-operated equipment or as economical alternatives to electrical motors. When the safety of the process plant depends on the reliable operation of a steam turbine, the turbine system should be designed to start up and reach operating speed in a way that minimizes hazards.

3.8.2.1. Types of Steam Turbine

Steam turbines are of two basic designs, impulse and reaction (or compound velocity). A simple single-stage impulse turbine consists of a wheel that resembles a rotating disk with a single row of buckets or vanes mounted on the circumference. One or more nozzles direct steam against the buckets, causing the wheel to rotate. The resulting rotation is transferred to the driven equipment. A multistage impulse steam turbine has a series of wheels connected to a common shaft with jets directed at each wheel on the shaft. A reaction turbine is a multistage machine with alternating rotating and stationary blades. Steam impinges on the blades of the rotating wheels. As the steam leaves the rotating blades, it enters a stationary blade set that reverses the direction of flow. This redirects the steam onto the next rotating blade set. The repeated redirection of steam flow between rotating and stationary blades, called velocity compounding, continues until the exhaust pressure is reached and the steam leaves the turbine.

3.8.2.2. Steam Turbine Operating Parameters

Safe operation of steam turbines begins with proper selection for the application. When a steam turbine is to be specified, the buyer should furnish, as a minimum, the following information:

Operating conditions
- Normal and maximum power requirements for the rotary equipment. This information permits the vendor to establish the rated power for the turbine.
- Service—continuous, intermittent, or emergency standby.
- Temperature, pressure, and maximum continuous supply rate of the available steam.
- Required turbine speed (rpm).

Location
- Exact location of plant—elevation (or actual atmospheric pressure); average and extreme summer and winter temperatures.
- If indoors, whether or not area is heated.
- If outdoors, whether or not the turbine is exposed to the elements or is under a roof.
- Turbine level—at grade or if not, indicate the need for vibration dampening.
- Any hazards in turbine area; exposure to dust or fumes. Selection of NEC class, division, and hazard group, if appropriate.

3.8.2.3. Steam Turbine Design Safety

Proper design of a steam turbine and associated controls and piping is of primary importance. The purchaser should confirm that the vendor has provided the following information:

- turbine type
  —horizontal or vertical
  —condensing or noncondensing
  —single- or multistage
  —steam exit conditions (for each stage)
- performance curves
- mechanical, electrical, instrumentation, and control drawings
- utility requirements
  —cooling water
  —electricity
  —lubrication
- materials of construction of all parts (API Standards 611 and 612 contain detailed information.)

The following points also should be checked:

- If a turbine is an alternative power source for rotating equipment, adequate controls should be designed to provide a satisfactory transition to turbine drive during a power failure, and an orderly return to the electric drive once power is restored.
- Sufficient boiler capacity should be available to prevent undesirable carryover in the steam.
- Water treatment system performance should be adequate [13].

3.8.2.4. Steam Turbine Installation Safety

Factors that must be considered in the installation of the selected system include:

- The turbine shipped to the site must match the vendor's drawings.
- The site must be ready for turbine installation; anchor points must be correctly installed.
- Proper erection equipment and qualified personnel must be available at the site.
- The turbine must be oriented correctly. The piping and electrical connections must conform to specification; variances must be documented and approved.
- Large turbines typically have horizontal split casings. Room should be available overhead for a hoist or traveling crane to lift the casing for maintenance. The height should also be adequate for the lifting of turbine components over the adjacent equipment. A monorail or crane should be located nearly over the center line of the turbine to minimize the possibility of injury or damage from manhandling of components. Room should also be left for laydown of dismantled components.
- Piping and supports must be installed to minimize stresses on the turbine.
- Piping and equipment must have proper thermal insulation.
- Vibration dampers must be installed as needed.
- The electrical installation must match the hazard designation of the area.
- The noise dampening system must be properly installed. Turbines can have noise levels around 100 dB. Noise propagation through a floor can be reduced by support on dampers.

3.8.2.5. Steam Turbine Operating Safety

Once the steam turbine is installed correctly, it is ready for performance testing and operation. Testing should follow the ANSI/ASME Test Code PTC 6-1976. After a pre-startup check, the turbine can be started and its operation monitored to determine whether any of the following problems occur:

- vibration over a wide operating range
  —dirty bearings
  —tight bearings

—bowed or out-of-balance shaft
—misalignment
—scale accumulation on blades
- damage due to presence of wet steam
- slow acceleration or failure to reach full speed
  —incorrect governor setting, causing steam to cut off too soon
  —load heavier than specified
  —steam line blocked, fouled, or undersized
  —back-pressure higher than specified
- erratic speed; governor continuously hunting
  —steam control valve sticking
  —governor's mechanical linkages need tightening
  —valve stem or governor spindle bent
- hot bearings
- oil cooling system not operational
  —cooler fouled
  —oil filter fouled
  —cooling water piping fouled or blocked
  —oil contaminated or emulsified
  —wrong oil used

3.8.2.6. Steam Turbine Maintenance Safety

A good maintenance program will ensure continued safe operation of a turbine [14]. Preventive maintenance will include systematic monitoring of the turbine's operation. For example, steam consumption may vary with time, suggesting a change in turbine efficiency. When this occurs, the cause should be identified. Some of the variables that should be monitored are:

- vibration history
- pressure and temperature at the turbine inlet, outlet, and interstages, for evidence of leaks or damaged insulation
- bearing oil pressure and temperature (trends may indicate fouling, bearing wear, or system leaks)
- steam system performance and absence of leaks at joints or valve steams
- throttle valve and governor operation
- power efficiency

When a turbine undergoes routine maintenance, the following checks should be performed:

- inspection requiring no major disassembly
  —thrust clearances
  —radial bearing clearances
  —alignment

—casing keys and baseplate
—piping supports and anchors
—Borescope inspection of interior

Inspection and overhaul requiring major disassembly should be incorporated in the preventive maintenance program.

3.8.3. Transmission

Simple apparatus may be connected directly to the shaft of a motor or a turbine. More often, there will be some form of transmission device between the driver and the load. This device may be a coupling, a gear train, a belt, or a chain. The following discussion will not cover variable-speed drives. Most of the safety considerations for the fixed-ratio devices treated below apply to all moving and rotary equipment.

3.8.3.1. Belt Drives

Belt drives are usually chosen for their simplicity and low cost. They are not positive drives, as are some of the alternatives discussed below, but will allow up to about 3% slippage (belt drives should therefore not be used with synchronous motors). Flat belts in particular have some limitations in speed and in their ability to transfer power.

V-belts are usually considered to give the best combination of performance and life. They take their name from their normally trapezoidal shape in which the narrow end is wedged into the pulleys. There are many variations on this shape (poly-V, double V, multiple V, link V, cogged V). The most popular construction is a jacketed rubber belt with reinforcing cords. The cords give the belt its strength and its ability to carry a load. Cords may be of steel, glass fiber, or polymeric material such as rayon.

Belts are rated for service by the Mechanical Power Transmission Association. Having determined that a belt is rated satisfactorily for the job in hand and that it can accept the maximum torque developed by the motor, the user should verify the following:

- guards installed to prevent contact of personnel with belt
- adequate clearances
- layout provides safe access during maintenance
- tension control adequate
- means of "takeup" provided to offset stretching of the belt
- belt speed and temperature within design limits
- heat dissipation adequate
- static dissipation provided when necessary (e.g., use of conductive belts with flammable materials)
- multiple belts used where failure of the drive would cause a hazard

3.8.3.2. Chain Drives

Chain drives are superior to belts in their ability to transfer power. The usual form is a metallic wraparound chain formed of links that can be riveted or cottered. Riveted construction is more common in smaller sizes.

Where chain drives are used, manufacturers' catalogs can be consulted for selection and for advice on alignment, lubrication, and maintenance. Chain drives should be manufactured in accordance with ANSI B29.1.

Most of the safety considerations listed above for belt drives are applicable to chain drives.

3.8.3.3. Gear Drives

Gear drives are favored for smooth, accurate transmission of power. This is important in larger units, but less so in drives typically found in small plant operations. Gears should be manufactured to AGMA Standard 390.03.

3.9. FILTERS

Filtration is the removal of solids from a fluid stream by capturing them on a solid medium which allows the clear fluid to pass through. Because of this operation's ancient origins, an extremely wide selection of apparatus is available; however, major advances have been made in some design concepts. Figure 3-16 summarizes the major classifications, gives their characteristics, and suggests criteria for selection.

A filter operates by one of two mechanisms. In cake filtration, the solids (perhaps including a filter aid) form a cake on the surface of the filter medium. Other important filtration steps include washing, to remove occluded liquor, and dewatering, to produce a cake of higher solids content. The other mechanism is depth filtration, in which the solids are captured within the filter medium itself. This might take the form of a bed of granular solids or a porous solid structure. Depth filtration is most often applied to dilute slurries.

3.9.1. Safety Considerations

Design and operating considerations include

- proper choice of materials of construction (Section 3.1)
- leakage and spill control (Section 2.4)
- containment of highly hazardous materials (Chapter 4)
- prevention of static electric discharge when handling flammable materials (Section 2.2)
- proper handling and control of unstable materials (Chapter 4)
- hygiene

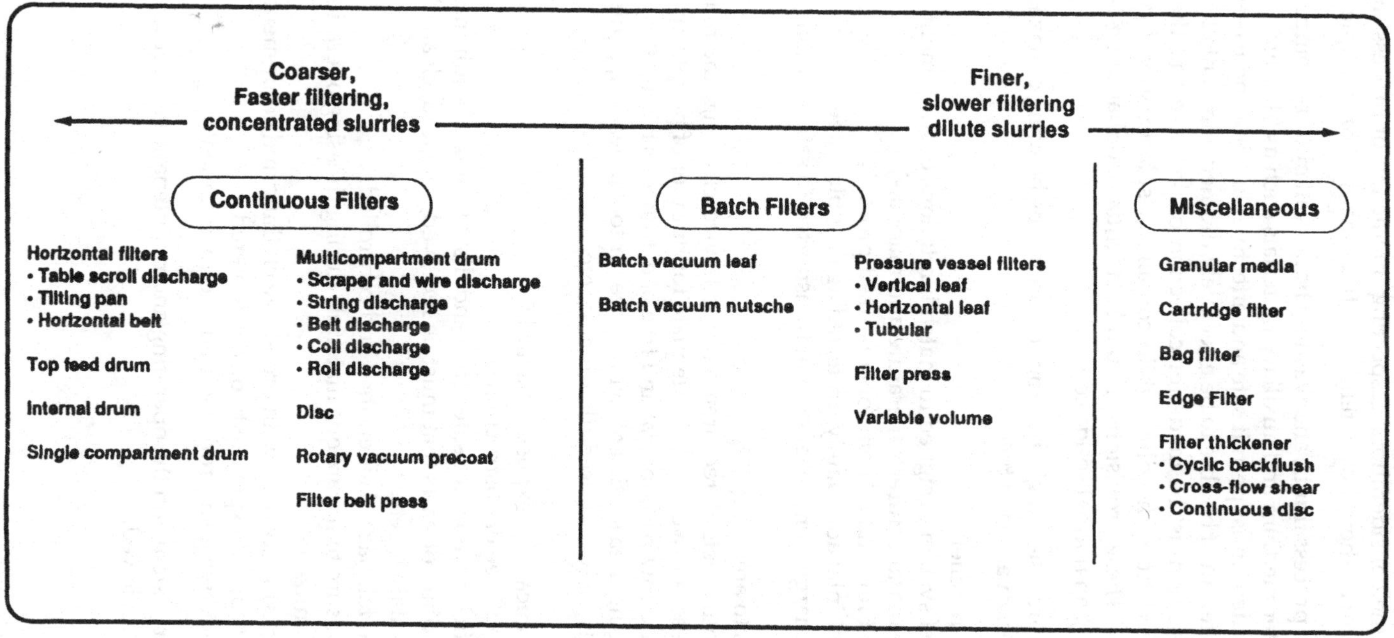

Coarser,
Faster filtering,
concentrated slurries
───────→

Finer,
slower filtering,
dilute slurries
───────→

Continuous Filters

Horizontal filters
· Table scroll discharge
· Tilting pan
· Horizontal belt

Top feed drum

Internal drum

Single compartment drum

Multicompartment drum
· Scraper and wire discharge
· String discharge
· Belt discharge
· Coil discharge
· Roll discharge

Disc

Rotary vacuum precoat

Filter belt press

Batch Filters

Batch vacuum leaf

Batch vacuum nutsche

Pressure vessel filters
· Vertical leaf
· Horizontal leaf
· Tubular

Filter press

Variable volume

Miscellaneous

Granular media

Cartridge filter

Bag filter

Edge Filter

Filter thickener
· Cyclic backflush
· Cross-flow shear
· Continuous disc

FIGURE 3-16. Types of Filters and Their Industrial Applications.

Some of the considerations respecting interaction of process fluids with their environment should be extended to filter media. For example:

- Solids, process fluids, and washing liquors all must be compatible with the filter medium and auxiliary materials such as filter aid.
- The system must permit safe and effective flushing and cake washing.
- The filter must be able to be effectively drained and vented.
- Provisions must be made for collection of the cake and the medium removed from the filter while containing splashing and controlling vapors. These provisions should be adequate for handling cakes with lower than normal solids content.

Other approaches might be grouped under the heading of hazard control by system design as listed below:

- pressure relief
  —relief system designed to match limitations of equipment (see manufacturer's literature and manuals)
  —relief line and all components sized properly
  —relief inlet adequately protected against blockage
  —discharge from relief system transported to safe location

- access to area
  —layout meets all requirements for service, cleaning, and maintenance
  —access and work space adequate for number of people involved
  —work flow path logical and free of extraneous hazards
  —staging area sufficient, with space for tools, cleaning equipment, parts, new and used filter media, etc.

- mechanical and exposure hazards
  —layout free of obstruction
  —sufficient room for safe setup and dismantling of equipment and for handling of parts that must be removed, replaced, cleaned, maintained, etc.
  —slip/fall hazards identified and controlled
  —exposure to hazardous materials during dressing operations minimized
  —workers provided with protective clothing and equipment adequate for types of exposure to be encountered
  —protective gear properly used and maintained

The filter system and the operating and maintenance procedures must be periodically reviewed.

3.9.2. Waste Minimization and Disposal

The purpose of filtration is the separation of solids from a fluid phase. The solids, the fluid, or both may be valuable. In most cases, however, one phase is considered to be a waste and its minimization and disposal are important parts of the process.

The first consideration should be waste minimization. This might involve a change in the process to eliminate the waste, recycle of a material previously wasted, recovery of a valuable or reusable component, reuse of wash liquors, or improved washing of a cake to reduce its hazardous nature. Waste minimization offers the following:

- improved economics
- lower disposal costs
- less generation and handling of hazardous materials
- more responsible and better utilization of resources

Even with waste minimization, however, filtration will usually produce some material that requires disposal. Proper handling requires the following:

- equipment and medium cleaning and decontamination
- collection and disposal of waste solutions
- identification of hazards and their communication to employees, waste disposal contractors, shippers, and the public
- adequate procedures and documentation
- proper materials of construction for handling equipment and containers
- recognition of any problem created by separation of phases during handling, shipment, or unloading of wastes
- tracking of wastes throughout the process of treatment and final disposal
- disposal in accord with all regulations, agreements, and principles of safety and product stewardship
- securing and monitoring of disposal site

3.10. CENTRIFUGES

Centrifugal machines are gravity intensifiers. One type causes separation between two phases of different densities when at least one of the phases is a liquid. Another type causes a liquid to flow through a solid. The former type might be considered a centrifugal settler and the latter a centrifugal filter. Figure 3-17 shows schematic diagrams of the basic types of centrifuges.

BATCH

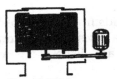

**Manual Discharge
Batch Centrifugal
Filter**

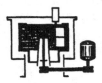

**Variable-Speed
Batch Automatic
Centrifugal Filter**

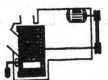

**Constant-Speed
Batch Automatic
Centrifugal Filter**

INTERMITTENT

**Scoll-Discharge
Centrifugal Filter**

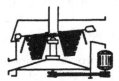

**Constant-Angle
Screen Bowl Filter**

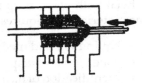

**Pusher-Discharge
Centrifugal Filter**

CONTINUOUS

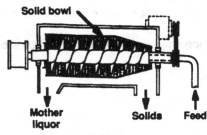

Solid bowl

Mother
liquor   Solids   Feed

**Scroll-Discharge
Centrifugal Settler**

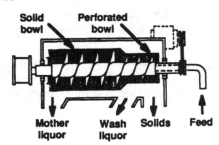

Solid   Perforated
bowl   bowl

Mother   Wash   Solids   Feed
liquor   liquor

**Scroll-Discharge
Centrifugal
Settler/Filter**

FIGURE 3-17. Types of Centrifuges.

Centrifuges are unique among the types of equipment covered in this book in their production of huge inertial forces during operation. This has several consequences:

- Live stresses on supports and on the building itself can be significant.
- Acceleration and deceleration must be gradual; there must be no "sudden stops."
- Alignment of large components and rotating parts is critical to safe operation and has very small tolerances.
- Careful balancing of machinery is essential, and vibration monitoring is often advisable.

A machine should not be purchased or installed without careful consideration of these factors. All personnel must recognize the mechanical hazards of a centrifuge and realize that inertia will keep the rotating element in motion for some time after the "Stop" button is pushed.

Proper installation, charging, and setting of key parts require specialized skill and a knowledge of the supplier's requirements for positioning and alignment. With certain models, manufacturers will supply jigs to help the user to set rotating components accurately.

3.10.1. Types of Centrifuges

3.10.1.1. Centrifugal Filters
Centrifugal filters are usually characterized as basket and screen centrifuges. All are capable of filtration, washing, and dewatering, in that sequence. Table 3-11 shows ranges of application of centrifugal filters. The most important variables are the concentration of the feed slurry and the size range of the solid particles.

3.10.1.2. Centrifugal Settlers
Centrifugal settlers are used to thicken or dewater solids; Table 3-12 gives their application range. The solid-bowl centrifuge is highly versatile and is probably the most familiar example of this type. It can handle wide ranges of slurry concentrations and particle sizes. Inside the rotating bowl is a conveyor with a differential rate of rotation that carries the concentrated solids out of the centrifuge. The solid-bowl machine allows some washing of the product.

TABLE 3-11
Application Range of Centrifugal Filters

Slurry Characteristic	Centrifuge Types						
	Continuous, Slip, Screen	Continuous, Vibratory, Screen	Continuous Conveyor	Single/ Multistage Pusher	Automatic, Batch-Plow, Horizontal	Automatic Batch-Plow, Vertical	Manual, Batch-Plow, Vertical
Minimum particle size*	250	500	150	40	20	20	10
Maximum particle size	10,000	10,000	5,000	5,000	2,000	1,000	1,000
Slurry solids, %	40–80	40–80	25–75	15–75	10–50	5–20	2–10
Filtering time, s/cm	0–3	0–3	1–5	1–10	3–20	10–30	20–60
Process Capabilities							
Solids rate, tons/h	5–40	5–150	1–150	0.5–50	0.25–20	0.1–5	0.1–1
Washing ability	Fair	Poor	Fair	Good	Good	Excellent	Excellent
Discharged-cake condition	Dry, granular	Dry, granular	Dry, granular	Dry, granular	Dry, granular	Firm, granular	Pasty, granular
Liquid clarity	Fair	Fair	Poor	Good	Good	Excellent	Excellent

*Minimum particle size (microns) for practical economical application.

3.10.2. Design Considerations

3.10.2.1. Maintenance and Cleanup Access
Batch centrifuges require frequent cleaning, modifications to their setup, and preventive maintenance. Safety and effectiveness in these procedures require sufficient space with convenient access and easy egress.

TABLE 3-12
Application Range of Centrifugal Settlers

Slurry Characteristic	Centrifuge Types					
	Solid Bowl	Nozzle-Disk	Wall-Valve Disk	Manual Disk	Manual Tubular	Manual Plow, Batch
Minimum particle size*	2	0.25	0.25	0.25	0.1	2
Maximum particle size	5000	50	200	200	200	5000
Slurry solids, %	2–60	2–20	0.1–5	<1.0	<0.1	0.1–5
Test spin-time, min.	0–3	1–10	1–10	1–10	2–20	0–3
Test sediment-condition	Firm paste, granular	Fluid	Flowable, pasty	Pasty, firm	Pasty, firm	Firm
Process Capabilities						
Solids rate, lb./hr	100–100,000	10–3,000	1–1,500	1–100	0.1–5	20–5,000
Liquid rate, gpm	1–500	1–800	1–200	1–500	0.25–20	1–100
Washing	Some	Some	No	No	No	No
Cake condition	Pasty, granular	Flowable, pasty	Flowable, pasty	Pasty, firm	Pasty, firm	Firm
Liquid clarity	Generally excellent	Generally excellent	Generally excellent	Generally excellent	Generally excellent	Generally excellent

*Minimum particle size (microns) for practical economical application.

3.10.2.2. Inerting Requirements

When handling materials that emit flammable, toxic, corrosive, or otherwise hazardous vapors, it may be necessary to fill a centrifuge with a blanket of inert gas. The inerting gas is chosen to suit the particular job; nitrogen fits most of these applications and is the most widely used inert gas in process plants.

When an inert gas is used, its supply must be secure and adequate in volume. Because some centrifuges act as fans, it is important that the user verify the adequacy of inert gas flow. This may require some initial experimentation and an occasional check of the centrifuge atmosphere during operation.

The inert gas, which usually is at least an asphyxiant, must not be allowed to become a hazard itself. It should be handled with all due precautions, as outlined in Section 2.6.

3.10.2.3. Venting

Venting is required in all cases. When a purge gas is applied, it often is intended to dilute some compound below a hazardous level or to provide positive removal of a vapor. The resulting gaseous mixtures must be collected for safe venting and possibly for treatment. The treatment process must suit the hazardous material. The system must be designed and installed to integrate treatment with the centrifuging operation. Piping, for example, should be direct, simple, and without low points that might trap liquid.

The ideal treatment process will accept vents directly from the centrifuge; if intermediate pressure boosting is necessary, the possible entry of air should be considered. If this were to cause a problem, a new design may become necessary.

3.10.2.4. Alarms and Shutdowns

Centrifuges require particular attention to their mechanical hazards. Certain types will develop forces several thousand times as great as normal gravity. An imbalance in loading can quickly create a dangerous situation. If such is the case, the machine should have an automatic stop device.

Swirling of liquid or slurry inside the centrifuge is one frequent cause of imbalance. Introducing the feed at the proper rate during acceleration is the best way to avoid this. Under no circumstances should feed begin before the rest of the system is ready to accept and process it. When the centrifuge is instrumented, interlocks are required.

When machines are opened routinely, as in batch operation, systems must be provided to prevent inadvertent starting. Another interlock should prevent operation when any normally closed apertures are open.

It is advisable to install various alarms, and some users or manufacturers will specify trouble alarms which warn of faults before shutdowns are activated. These should be used only if they give an opportunity to forestall the shutdown.

As an example, a typical basket centrifuge installation requiring a nitrogen blanketing system would need the following alarms and shutdowns:

- low nitrogen flow or supply pressure
- high oxygen content in vent
- high/low vent line pressure
- high motor temperature
- high bearing temperature

- high/low basket speed
- excessive vibration
- open door
- closed discharge gate
- high level in discharge chute
- advanced/retracted scraper
- excessive cake thickness.

3.10.3. Operation

Safe operation of centrifuges requires:

- detailed written operating procedures
- operator training
- operator's conformity with procedures
- adequate maintenance
- proper collection and handling of products, liquors, wastes, and vents

Operating procedures should be written, validated, and periodically audited. The audit will also serve as a check on the adequacy of the operator training program. The operator of a centrifuge must be familiar with all the hazards of its operation and have a thorough working knowledge of the safety systems.

Maintenance procedures and frequencies are equally important. The supplier's recommendations should be followed carefully. Preventive maintenance should be emphasized.

3.10.4. Waste Minimization and Disposal

Wastes may be solids, liquids, or gases. A centrifuge produces a sludge or cake containing the solids that were removed from suspension. If this sludge is a waste product, the first consideration should be to reduce its quantity. Next, a safe method must be designed for its disposal. This involves the same considerations as does the disposal of a filter cake (Section 3.9) or other solid wastes (Section 2.7).

Alternatively, the liquid removed in the centrifuge may be the waste material. The section on filter waste minimization and disposal (3.9.2) lists the relevant safety precautions.

In addition, the purge gas may become a waste that requires proper handling.

3.11. DRYING AND PARTICLE SIZE REDUCTION

These two quite different operations share the special hazards that are presented by small, dry particles. These hazards, which relate to surface activity, include static electricity accumulation and the possibility of dust explosions. These can occur in any type of apparatus. No product, therefore, should be dried, ground, or milled without thorough knowledge of its safety-related properties. All of the following must be considered in a hazard evaluation:

- burning behavior
- dust explosion behavior
- ignition temperatures (dry product and any flammable liquid in the feed)
- thermal and impact stability

Approaches to be considered in process design and equipment selection include:

- limiting the temperature of a dryer's heating medium (e.g., by thermal cutout switches or steam pressure reducing valves)
- blanketing with an inert gas
- conforming to requirements of the appropriate electrical classification
- suppression of explosions
- installation of a fire-protection system that satisfies all codes and insurance requirements
- use of a deluge system
- safeguarding against loss of a utility or vacuum (e.g., inerting, cooling, venting)
- when charging or discharging material, protecting against:
  —mechanical hazards
  —vapor emissions
  —loss of inert gas
  —accumulation of hazardous dusts or gases
  —accumulation of static electricity
- protecting against static discharge in flammable vapor atmospheres (e.g., control of process, inerting, use of conductive internals, grounding)

All applicable factors must be considered for each different product and each type of equipment. A review is necessary whenever there have been irregularities during production, changes in raw material, or modifications to the process or the equipment.

These general topics apply to a broad range of apparatus. The sections that follow give some equipment-specific considerations.

3.11.1. Dryers

Figure 3-18 lists some of the many different types of equipment used for drying. Because of some basic differences, they will be discussed in more detail than some of the equipment covered in earlier sections.

Before considering drying as a method of removing solvent from a solid material or selecting specific drying equipment, the material should be tested sufficiently to determine whether it can be dried safely. The tests should measure

- flammability of the solvent
  —Determine the explosion limits, flash point, and autoignition temperature of the solvent.
  —Determine whether drying can be performed at operating conditions that fall within a safe range.
  —Determine whether inerting or vacuum operation is feasible.
  —Determine whether the process can become exothermic. If tests show, for example, that an exotherm occurs above 140°F, perform the drying well below this temperature.
- reactivity or flammability of the material being dried
  —Determine whether the material decomposes on heating and whether the by-product formed is toxic, flammable, or explosive.
  —Determine the temperature above which by-product formation begins.
  —Determine whether the material is self-heating. Some organic materials, such as peroxides and nitrogen compounds, may decompose when they are dried. The reaction in some cases can be rapid and explosive. The testing should determine whether the reactivity of the material is classified as nonexothermic, exothermic but not runaway, or runaway.
  —Determine whether the reactivity can be reduced or eliminated by inerting, using vacuum, or operating at lower temperature.
- static electricity produced by tumbling, aerating, or atomizing the material during the drying process
- dust production capability of the process and the hazardous properties of the dust

The provision of thermal energy to a dryer is a separate consideration. Heating may be indirect, for example by steam or by the use of waste heat from another processing step. It may also be by direct firing, in which case proper handling of the feed and design of the combustion system are essential.

There are many references and design codes dealing with combustion safeguards. Factory Mutual's Loss Prevention Data Book Service publishes data sheets that describe safeguards against fires and the prevention of fuel explosions. Examples are number 6-9, Industrial Ovens and Dryers, and numbers 6-17/13-20, Rotary Kilns and Dryers. The NFPA also has a series of

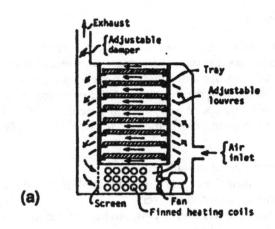

(a)

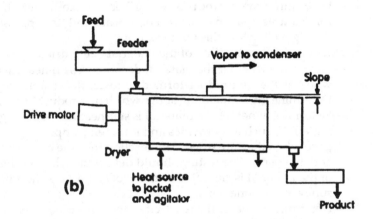

(b)

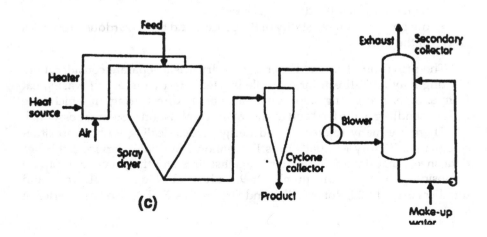

(c)

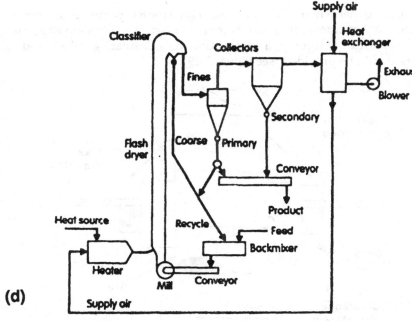

(d)

FIGURE 3-18. Typical Drying Systems. (a) Batch atmospheric tray dryer; (b) horizontal rotary kiln dryer using indirect heating; (c) direct-contact spray dryer; (d) direct drying system with backmixing, milling, classifying, product recycle and heat recovery.

standards dealing with combustion systems, ovens, and furnaces. Some of these are aimed at very large systems, but the principles apply also to the equipment usually found in process plants.

In summary, energy supply systems must be built with regard to their safe operation while exposed to an inherent fire hazard. Prime considerations include:

- choice of location
- provision of sprinklers or deluge systems
- location relative to other equipment and materials stores
- use of noncombustible construction throughout the installation

Fuel or vapor explosion hazards require installation of explosion vents. These can be blowout panels that also provide some insulating value and structural strength. A frequent practice is to provide a vent ratio of one square foot of surface to every fifteen cubic feet of dryer volume. This is illustrative only, and may serve as a first estimate. Venting systems must conform to NFPA 68.

Where high temperatures are required for drying, stack temperatures may be high enough to require protective measures. Ducts should be insulated. The space around them should be ventilated to keep the temperatures of surrounding structural components low. If there is no reasonable alternative to passing a hot exhaust duct through a combustible roof, insulating collars and rain shields should be provided (see Figure 3-19).

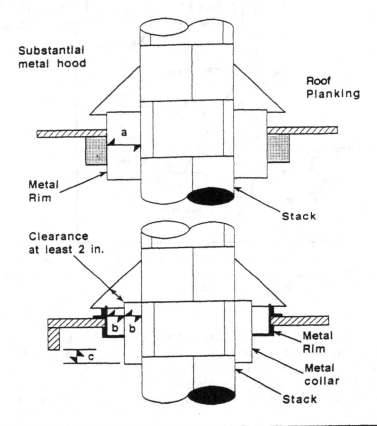

Stack temp.	Stack diameter	Clearance, in.		
		a	b	c
Under 600°F	8 in. or less	8	2.5	6
	More than 8 in.	1 2	4	9
600°F and over	8 in. or less	1 2	4	9
	More than 8 in.	1 8	6	1 2

FIGURE 3-19. Separation of Stacks from Combustible Roofs.

3.11.1.1. Drying Ovens

The basic oven is a simple chamber into which wet or dissolved solids are placed and then dried by the application of heat. The energy-transfer medium may be a hot surface or a gas. The chamber may simply be an enclosed space, or it may contain receptacles for the material to be dried.

Most ovens found in smaller operations are batch models. In one respect, batch ovens are more hazardous than continuous ovens because there will be a peak in the rate of evolution of volatile material. In a continuous process, there is a nearly constant, substantially lower rate of evolution. Much lower rates of ventilation may suffice for dilution of this vapor. In a batch process without complex control or close monitoring, the ventilation rate during much of the cycle will usually be much higher than ideally required. This is wasteful of ventilating gas and thermal energy, and it makes the task of safe removal and recovery of solvent vapors more difficult.

Where vapor explosions can occur, prevention is possible by careful monitoring of the process, the study of records, or appropriate instrumentation and real-time observation. Examples are:

- frequent problems in a particular oven
- slow heatup, which creates the temptation to cut the safety ventilation rate
- incomplete drying of charge in standard time
- condensate deposits in stacks or around oven doors

Direct observations that may indicate problems include:

- cold exhaust system due to reverse flow/inadequate ventilation
- negative pressure in oven or room relative to surroundings
- instrumented measurements out of normal range

When the process or product is changed, a review of the ventilation requirement is necessary.

Other safety features that may be applied to drying ovens are:

Purging
After any interruption in operation, an oven should be purged to remove flammable vapors. Purging should be automatic before a restart from a heating or ignition system shutdown. When fresh air is used, there should be at least three changes of the oven atmosphere.

Internal circulation
Induced circulation within the oven will prevent accumulation of high vapor concentrations in dead areas.

Vent duct design
Ventilation-control dampers should not be able to stop the flow of air completely. Dampers should have positive mechanical stops or be cut away in order to leave some space in the cross-section of a duct at the line of closure.

The exhaust duct from the oven should be placed to minimize short-circuiting and thus to avoid accumulation of vapor in pockets. A safety vent should run outdoors as directly as possible, and the building ventilation system should be designed to prevent reentry. When it is necessary to run vents from more than one source to a recovery or destruction process, the design should account for any reasonable combinations of flow.

Interlocks
Interlocks should ensure that all vent and circulating fans are on before feed begins. Fan failure should stop feed, deactivate ignition, and close the appropriate valves. Feed must be on before safety shutoff valves are opened and before the heating system can be activated.

3.11.1.2. Vacuum Dryers
A ventilating gas will increase the rate of evaporation from an oven by reducing the partial pressure of the volatile component. Vacuum ovens achieve the same effect by reducing the total pressure in the system. Using vacuum, however, does not preclude the simultaneous use of a ventilating gas. The associated hazards then are the same as those discussed in the previous section.

Even when external gas is not supplied, vacuum dryers present the added problem of inleakage of air. This can be offset by proper seal design, the use of balanced-gas seals on rotating openings, and standard combustion safeguards [15].

Many vacuum dryers use steam-jet ejectors with water-cooled after-condensers. These conditions create a concern about intrusion of water back into the dryer if vacuum is lost. This is particularly important if water-reactive products are present.

3.11.1.3. Fluidized-Bed Dryers
In fluidized-bed units, drying gas flows through a bed of the material to be dried at a rate sufficient for fluidization. With some products, these dryers can achieve very high drying rates or, conversely, operate at relatively low temperature. They often consume less energy and require less floor space than other dryers.

In the typical fluidized-bed dryer, the drying gas is hot air. Air is pulled over heating coils and through the dryer by an induced-draft fan. A particular hazard here is the possibility of dust explosions. Fine particles, still in a fairly dense phase, are surrounded by warm or hot air. The hazard is more severe if flammable solvents are being removed. A synergistic effect may exist between dusts and flammable vapors. Even at vapor concentrations below the LEL, the ignition energy of the mixture may be considerably lower than that of the dust alone. This emphasizes the need for the precautions listed for other dryers and in the section on dust explosions (Chapter 2).

One alternative is to blanket the dryer with an inert gas. Because of the operating expense of an open-loop system, this usually involves recycle and solvent vapor recovery from the drying gas. These steps add to both capital and operating costs of the fluidized-bed dryer and offset some of its advantage in energy and floor space requirements.

3.11.1.4. Spray Dryers

In spray dryers, a solution or dispersion of solids is subdivided into a fine spray to assist the evaporation of the liquid into a hot gas. The spray usually is produced by a nozzle (by pressure or an atomizing gas) or a spinning disc. Flow usually is downward, and the dried product is swept out of the chamber with the gas.

Process-side hazards arise from the suspension of a combustible powder in air and from the accumulation of the product in the dryer or ductwork. Major safety precautions are

- temperature control of the hot gas
- provision of high-temperature alarms and shutdowns
- explosion venting
- fire protection

Some larger dryers use air-diffuser plates in their upper sections to improve distribution. Relief in such a case must be from a lower level, in the body of the dryer.

An external fire protection system may have to be adapted to the geometry of a spray dryer. Sprinklers should be applied directly to the dryer, on the head or below the air diffuser plate. They may also be required below a supporting skirt. The temperature ratings of such sprinklers should be appropriate to their locations and the operating temperature of the equipment.

3.11.1.5. Rotary Conical Dryers

The rotary conical dryer blends as well as dries. Certain safety requirements apply especially with this type of drying equipment, such as:

- An effective ventilation system in the dryer area to protect personnel from hazardous dust and vapors when the dryer is opened for charging or discharging.
- Operating procedures and personnel training in loading and discharging material. The procedures and training should emphasize minimizing exposure to hazardous material.
- A cage surrounding the dryer to prevent personnel injury due to improper entry of the area while the dryer is tumbling. A door should be supplied with locking and interlocks to prevent entry while the dryer is tumbling and to prevent startup of the dryer if the cage door is not properly latched. The dryer operating panel should include door status lights.

- Equipment for grounding the dryer to prevent fire and explosions caused by the buildup of static electricity.
- Vacuum protection. Alarms should be provided on the dryer operating panel indicating possible air in-leakage.
- Inerting for those dryers required to operate where solvent vapors are present. Alarms should be provided on the dryer operating panel indicating possible air in-leakage or loss of inert gas (e.g., oxygen measurement).
- Maintenance procedures, preventive maintenance schedule and personnel training for maintaining the dryer and its operating area in a safe condition. Maintenance should stress:
  —door gaskets
  —seals between steam piping and tumbler jacket, and between vapor exhaust piping and tumbler interior
  —elimination of solids and dust accumulation

3.11.2. Size Reduction Equipment

The range of applications of size reduction is very wide. It includes the preliminary breakup of large masses into pieces that can be handled in a process and the grinding of smaller particles into fine powders. The size and design of equipment reflect this situation. Broadly, size reduction apparatus can be divided into those types which depend on mechanical crushing and those which use impact to fracture the solids. Table 3-13 shows some of the common types of equipment with their principal characteristics. Major hazards are those associated with the machinery and with the material being processed.

3.11.2.1. Mechanical Hazards

Size reduction usually requires a great deal of energy. Mechanical apparatus can seriously harm personnel in a very short time. Vendors should provide information on specific machinery, and interlocks can prevent operation when equipment is open or safety guards removed. Drivers and transmission systems will be robust, and it is important to observe the precautions described in Section 3.8.

The typical mill or grinder will be in a nearly closed system with obvious dangers. Few employees will think of approaching too closely or reaching inside the apparatus. The feed system often lacks secure cover and is not so imposing a hazard. The hazardous section should be clearly identified and posted with warning signs. Other useful precautions include:

- providing clear instructions on how to proceed when feed stops
- interlocking systems to shut down the size reducer whenever the feed system stops, and *vice versa*

TABLE 3-13
Classes of Size-Reduction Equipment

Type of Mill	Features	Operating Characteristics
IMPACT		
Hammer Mills	Hinged hammers attached to horizontal shafts.	High productivity; coarse product.
Vibrating Mills	Vibration of confined grinding medium inside cylinder or torus.	Uniform fine grinding. Low capacity; feed size must be quite small.
Rotating Cylinders	Ball mills, with metallic balls of uniform or graded size. Pebble mills, with ceramic balls. Rod mills, with metal rods nearly as long as the cylinder. Tube mills, with longer cylinders sometimes divided into compartments.	Low productivity; versatile. Can handle highly abrasive material (with increased maintenance cost).
Fluid-Energy Mills	Grinding by high-velocity gas streams.	Fine grinding; expensive equipment. Some control of product temperature.
COMPRESSION		
Jaw Crushers	Fixed gap between crushing surfaces.	Wide application, including large particles and abrasive matter.
Gyratory Crushers	Eccentrically rotating solid-cone impellers.	High capacity; usually finer grinding than jaw crushers.
Roller Mills	Solids crushed between rotating cylinder and wall of second cylinder.	Variable gap. Narrow product size distribution but limited size reduction ratio.

- never attempting to clear a blockage at the inlet to an operating size reducer manually or with the aid of a hand-held device
- avoiding work in the vicinity of moving parts
- not depending on the protection offered by equipment guards
- avoiding or protecting against entanglement of loose clothing, ties, long hair, etc.

The combination of high-energy rotating components or a high-velocity grinding medium with uneven demand (variations in feed rate and size distribution) can produce unusual levels of noise and dynamic structural load. Protection against the hazards of noise is covered in Section 4.8.3. Dynamic loads must be considered during design of a new facility. A competent structural engineer should review any addition of new equipment to an existing building.

3.11.2.2. Material Hazards

The previous section discussed the mechanical hazards associated with operation of size-reduction equipment. There will also be hazards related to the feed material or product. These hazards can be associated with the solid product itself or with impurities, residual solvent, or the like. Several problems are covered below in separate sections:

3.11.4. Packaging Hot Materials
3.11.5. Deflagration Hazards
3.11.6. Environmental Considerations

Table 3-14 lists other hazards and possible system failures, along with protective measures that can be taken to offset them. Other sections of this book cover many of these topics.

3.11.3. Screening Equipment

Screens are used in many different applications to separate solids according to particle size. As is true of other highly empirical operations, there are many different approaches to screening. The wide variety of machinery available is described in standard texts and vendors' literature.

Screens do not usually present material exposure hazards not already covered in Section 3.11.2. Generation of vapors is small, and dust control, while important, is not fundamentally difficult. Simple enclosures are often used to contain dusts. These range from partial enclosure of a vibrating element to total dust-tight enclosure of an entire system.

Proper installation of screens requires particular care because of the mechanical hazards involved and the dynamic load placed on the supporting structure. The latter is especially important when screening is integrated with size reduction or when resonant vibrations may be set up in the structure. The following are important safety considerations:

- careful handling and lifting to avoid excessive stress on members
- proper installation, closely following manufacturer's instructions (e.g., full-length support of screener)
- accurate alignment and balancing (proper setting of springs)
- proper setting of feed point
- access for operation and maintenance
- clearances for removal of equipment and changing of screens
- clearance between vibrating frame and adjacent supports
- noise control

Some devices that may be used to promote safe operation are:

- emergency stop button or wire at operating level and, if appropriate, in vicinity of screener at floor level

TABLE 3-14
Material Hazards and Protective Measures in
Size-Reduction Processes

Particle Hazards

Release to Atmosphere (See 3.11.6)
- Toxic
- Irritant
- Environmentally unacceptable

Surface Effects (See 3.11.5)
- Burning
- Ignition
- Dust explosion

Solvent Hazards (See 2.2)

Flammability

Thermal instability

System Failures

Loss of vacuum

Loss of inert atmosphere
- During process
- Charging/discharging

Protective Measures

Limit temperature of heating medium

Blanket system with inert gas

Area electrical classification

Electrical grounding (particular attention to removable internals)

Use of conductive cloth in dust collectors

Explosion-venting systems

Water deluge systems

- interlocks to prevent feed to screener when drive motor is not in operation
- grounding of machine body and screens when static electricity is a hazard

3.11.4. Packaging of Hot Materials

As drying or size reduction increases particle reactivity, some exothermic process may begin. Inside the apparatus, this energy may easily be dissipated by a surface or a fluid that acts as a heat sink. The exothermic process may continue when a product is gathered into a compact mass, as in a container. Energy dissipation then becomes much slower. Even a very slow and previously undetectable exothermic process can then begin to heat the mass. The rate of reaction will increase, and the final result may be autoignition of the product and burning or explosion of the container. This can occur many hours or even days after packaging.

This autoignition hazard can be avoided by evaluating the thermal stability of the product. Heat accumulation testing will define a maximum temperature for packaging. This temperature can be a function of the mass and shape of the material tested, and limiting container size is one means of controlling the hazard. If processing of the solid involves elevated temperature, it may be a simple matter to cool it before discharge.

When there is doubt about the thermal stability of a product or the possibility of heat buildup, it may be wise to quarantine the packaged goods for some time in a safe area.

3.11.5. Deflagration Hazards

Propagation of a combustion reaction can result in detonation or deflagration. A deflagration, with its lower flame speed, is the lesser hazard. These reactions can occur in any flammable medium, including dusts, and can be prevented or mitigated by techniques mentioned elsewhere in this volume and spelled out in the extensive NFPA publications and the loss-control literature [16, 17].

With certain materials, a locally initiated decomposition can propagate spontaneously to a deflagration even in the absence of oxygen. These materials include a number of fine, dry powders. This section deals with such hazards. Important points to note are:

- Deflagration cannot be prevented by inerting or operating under vacuum.
- Once initiated, deflagration cannot be stopped with an inert gas.
- Decomposition may produce large quantities of combustible gases.
- Many deflagrated residues are still combustible; a deflagration can change into a fire when air is admitted.
- Similarly, combustion of a deflagratable product can change into a deflagration even if deprived of air.

Materials to be dried or ground should be tested for deflagratability. Suppliers' MSDSs or standard references will contain this information when tests have already been made. Materials especially likely to deflagrate include diazo compounds, aromatics with nitro groups, and heterocyclics containing oxygen.

Deflagration hazards can be controlled in many different ways. These can be classified into three groups:

- modifying the product to reduce the likelihood of deflagration
- designing the process or the apparatus to reduce the hazard
- emergency intervention to mitigate consequences

The first method is to be preferred if the modification does not detract from use of the product. As noted above, drying in itself can increase the reactivity and sensitivity of a product. Controlled partial drying, achieved by a change in process conditions, may be suitable for the intended use while greatly reducing the deflagration hazard. Another approach is to add an inhibitor to the material. This will interrupt propagation of the chain reaction that leads to deflagration.

A second approach is to design the process to offset the possibility of deflagration. Intense localized heat input, for example, can be avoided by using equipment without moving parts in the processing zone. The absence of frictional heat often will allow normal control of the bulk temperature to be sufficient protection. Some equipment operates with only a small inventory of material. This not only reduces the severity should deflagration begin but also significantly reduces its probability through better control of process conditions, avoidance of a "critical mass," and faster dissipation of energy from the solid.

Finally, emergency systems can be installed to respond to the event and thus prevent a catastrophe. A deflagration is by definition more difficult to relieve than is a routine overpressurization and any relief system must be designed with this in mind. Early detection of the process and sizing for high rates of release are possible approaches. The deflagration process might also be interrupted. Since air oxidation is not essential to deflagration once the process begins, the use of inert gas is not a reliable means of prevention. Thorough wetting of the solids will often stop the reaction. This method requires a water deluge system, which with some products can be made more effective by the addition of a wetting agent.

3.11.6. Environmental Concerns and Hygiene

In any solids-processing operation, dust control is an important consideration. Where the solid product is conveyed from the apparatus, as with an air mill, highly efficient particle collection is necessary. Combinations of cyclones and bag collectors are often used in this application. Even when recovery is uneconomic, bag collectors may be used to prevent the nuisance of dust release. Other apparatus that does not recover the solids in an immediately useful form may also be used. This category includes wet scrubbers (e.g., venturis and various scrubbing columns) and electrostatic precipitators.

Dryers and some size reduction systems release solvents from the apparatus. With the possible exception of water, these solvents must not be allowed to escape into the environment. In most cases, the bulk of the evaporated material can be recovered by simple condensation or absorption. Such recovery processes often must be supplemented to reduce emissions to an environmentally acceptable level. Supplemental processes include low-temperature condensation (–60°C is often used for low-melting liquids), adsorption (activated carbon has wide utility), and chemical treatment (e.g., oxidation or contact with a specific reagent).

Many plants produce a variety of products and will frequently change from one to another. The loss of product and auxiliary materials should be minimized in these cases. Chapter 5 treats the subject of process changeover in detail.

3.12. INSTRUMENTS AND CONTROLS

The original purpose of installing instrumentation and control (I&C) systems in processing plants was to increase productivity by maintaining operation within established parameters with little or no human intervention. It was assumed that eliminating the human factor would eliminate the chance of operator error. As process plants increased in complexity, more sophisticated control systems were required to maintain productivity. The demands placed on I&C systems thereby increased. This created new and sometimes unanticipated reliability and safety problems.

With such rapid changes in I&C techniques, procuring the right equipment, installing it, and operating and maintaining it correctly present a challenge to the designer and process operator. The Instrument Society of America (ISA) has developed a set of standards and recommended practices that facilitate communication and establish norms for instrumentation and control systems. Table 3-15 cross-references various ISA documents as they apply to specification, installation, operation, and maintenance; Table 3-16 describes these documents.

Understanding the terminology and documentation is very important. ISA S51.1 is a useful general reference; ISA S5 explains the types of drawings used to describe the control systems; and ISA S20 gives examples of the standard forms used to specify instrumentation and controls. These three standards serve as a basis for well designed and safe process control systems.

3.12.1. I & C Design Safety

The first step in designing a safe instrumentation and control system is to consider three basic points:

1. Which process parameters will be
   - monitored (indicated)
   - logged (recorded)
   - controlled

TABLE 3-15
ISA Document Cross References

List of selected ISA Standards (S) and Recommended Practices (RP) related to process plant safety, cross-referenced to the activities of specification, installation, operation, and maintenance.

Document Number	Specifications	Installation	Operation	Maintenance
RP7.1	—	✔	—	✔
RP12.1	✔	✔	✔	✔
RP12.6	✔	✔	—	—
RP12.10	✔	✔	—	—
RP16.1,2,3	✔	—	✔	—
RP16.4	✔	—	—	—
RP16.5	—	✔	✔	✔
RP16.6	✔	—	✔	✔
RP31.1	✔	—	✔	✔
RP42.1	✔	—	—	✔
RP52.1	—	—	—	✔
RP55.1	—	✔	—	✔
RP60.1	✔	—	—	—
RP60.3	✔	—	—	—
RP60.4	✔	—	✔	✔
RP60.6	✔	—	—	—
RP60.8	✔	—	—	—
RP60.9	✔	—	—	—
S5	✔	✔	✔	✔
S12.4	✔	✔	✔	—
S12.10	✔	—	—	—
S12.12	✔	—	—	—
S12.13	✔	✔	—	✔
S18.1	✔	✔	✔	—
S20	✔	—	—	✔
S26	—	✔	—	✔
S50.1	✔	✔	—	✔
S51.1	✔	✔	—	✔
S71.01	✔	✔	✔	✔
S71.04	✔	✔	✔	✔
S75	✔	✔	✔	✔
S82	—	✔	—	—

TABLE 3-16
Description of ISA Documents: Recommended Practices (RP), Standards (S)

RP7.1	Pneumatic Control Circuit Pressure Test.
RP12.1	Electrical Instruments in Hazardous Atmospheres.
RP12.6	Installation of Intrinsically Safe Systems for Hazardous (Classified) Locations.
RP16. 1,2,3	Terminology, Dimensions, and Safety Practices for Indicating Variable Meters (Rotameters, Glass Tube, Metal Tube, Extrusion Type Glass Tube).
RP16.4	Nomenclature and Terminology for Extension Type Variable Area Meters (Rotameters).
RP16.5	Installation, Operation, Maintenance Instructions for Glass Tube Variable Area Meters (Rotameters).
RP16.6	Methods and Equipment for Calibration of Variable Area Meters (Rotameters).
RP31.1	Specification, Installation, and Calibration of Turbine Flowmeters.
RP42.1	Nomenclature for Instrument Tube Fittings.
RP52.1	Recommended Environments for Standards Laboratory.
RP55.1	Hardware Testing of Digital Process Computers.
RP60.1	Control Center Facilities.
RP60.3	Human Factors Engineering for Control Centers.
RP60.4	Documentation for Control Centers.
RP60.6	Nameplates, Labels, and Tags for Control Centers.
RP60.8	Electrical Guide for Control Centers.
RP60.9	Piping Guide for Control Centers.
S5	Various standards on the symbology and the types of design drawings used to describe the control process design and installation. Standards S5.3 and S5.5 cover process computer control display graphics.
S12.4	Instrumentation Purging for Reduction of Hazard Area Classification.
S12.10	Area Classification in Hazardous (Classified) Dust Locations.
S12.12	Electrical Equipment for Use in Class I, Division 2 Hazardous Classified Locations.
S12.13	Part I—Performance Requirements, Combustible Gas Detectors. Part II—Installation, Operation, and Maintenance of Combustible Gas Detectors.
S18.1	Annunciator Sequences and Specifications.
S20	Specification Forms for Process Measurement and Control Instrument, Primary Elements and Control Valves.
S26	Dynamic Response Testing of Process Control Instrumentation.
S37.1	Electrical Transducer Nomenclature and Terminology.
S50.1	Compatibility of Analog Signals for Electronic Industrial Process Instruments.
S51.1	Process Instrumentation Terminology.
S71.01	Environmental Conditions for Process Measurement and Control Systems; Temperature and Humidity.
S71.04	Same as S71.01 for Airborne Contaminants.
S75	Various standards on the design, specification, installation, operation, and maintenance of control valves.
S82	Various safety standards for electrical and electronic test, measuring, controlling, and related equipment.

2. Type of instrumentation system
   - self-contained, local instrument package
   - pneumatic transmitter/controller system
   - electronic system

3. Location of each instrument/control device (see ISA S71)
   - outdoors and exposed
   - outdoors and sheltered
   - sheltered and temperature controlled
   - sheltered with environmental control
   - hazardous environment

3.12.1.1. System Characteristics
A. System Specification
The specifications for instrumentation and controls should be complete and concise. Specifications should clearly define the design criteria listed above and should include the following:

1. Data sheets that list all the information required for design, installation, operation, and maintenance. To reduce the chance of any misunderstanding, standardized data sheets such as those described in ISA S20 should be used. Data sheets are completed partly by the purchaser and partly by the instrument/control vendor.

2. A narrative portion that provides information not given on the data sheet, such as:
   - plant name, address, and location (if different from main office)
   - scope of supply; i.e., what is to be supplied by purchaser and what is to be supplied by vendor.
   - list of vendor documentation to be submitted with the equipment:
     –installation instructions
     –operating manuals
     –maintenance manuals
     –drawings
     –calibration and shop testing data
     –code certificates, as required
   - testing and inspection requirements
   - list of recommended spare parts
   - list of recommended special tools
   - commercial terms and conditions
   - shipping and handling instructions

3. Special conditions that may affect any portion of the specification described above. Some conditions that may affect operational safety are
   - environmental effects
   - safety systems and the design of failure modes

- safety-related operator interfaces
- security configurations

B. Safety Systems

When considering safety interlock systems for a process plant, the designer should:

1. Before investing in a sophisticated safety control system, adopt an inherent safety philosophy [18] and in sequence consider
   (a) replacement of the hazardous material or condition by one that is less hazardous
   (b) reduction of a material hazard by
      •cooling it to make it less volatile
      •mixing it with another material to make it more inert
      •diluting it
   (c) reduction of quantities of hazardous materials in storage and in process
2. Avoid the use of safety interlocks to allow shortcuts in design or installation, use of less qualified personnel, or cutbacks in inspection and maintenance.
3. Consider whether safety interlocks are to be segregated on the system independently of the normal control system.
4. Analyze the failure modes of each final controller, valve, solenoid, relay, etc.
5. Provide redundant hardware at different levels, if necessary, to bring the system back to a safe condition.
6. Determine if a fault-tolerant control system is necessary. This is one in which a single fault can not cause the system to fail. Typically such a system uses triplicate redundancy with two-out-of-three voting logic and redundant functions.
7. Determine if an emergency power supply should be provided for any part of the system.

Traditionally, automated safety systems were contained within the single control system that also handled the process itself. More recently, there has been a tendency to provide dedicated controls for some of the more specialized safety devices. For example, a distributed control system (DCS) may be used for normal operation, startup, and shutdown, whereas a separate hard-wired interlock system with relay logic, possibly driven by a programmable logic controller (PLC), is used for emergency shutdowns. Safety systems are generally designed to operate during emergency shutdown conditions and for fire and gas monitoring and control.

Emergency Shutdown. An emergency shutdown system (ESS) provides a safe, orderly shutdown of a plant without operator action. The ESS has dedicated I/O and logical controls. The ESS and its components almost invariably receive power from a source with emergency back-up (e.g., a diesel generator).

The ESS operation is passive most of the time. It monitors potentially unsafe conditions and responds to problems by shutting down some part of the plant. To maintain the low probability of failure on which the plant design is based, proof testing of the ESS is necessary. This should be carried out on a scheduled basis and carefully documented.

Fire and Gas Monitoring and Control. The last monitoring and control level handles hazards due to combustible or toxic gases. It may communicate with the process control system. Again, it is usually a passive system and requires periodic proof testing. When the response to an alarm includes partial evacuation of the site or activation of response teams, frequent drills and clear establishment of responsibility are necessary.

C. Security Configuration
While good design, installation, operation, inspection, and maintenance are important, the control system must also be protected from tampering and inadvertent changes that can cause unsafe operation. Key points to be studied include:

- location of buttons and switches that can be operated accidentally or by unauthorized personnel
- ability of unauthorized personnel to accidentally or deliberately change or erase control software
- location of controls that can be by-passed without authorization

These vulnerable points usually can be secured by

- Design changes
  —Eliminate the vulnerable item if possible.
  —Relocate the item to an enclosure with a lock that can be opened only by authorized personnel.
  —Allow critical switches, buttons, and setpoint controllers to be operated only by a key restricted to authorized personnel.
  —Install software on computer control systems and security hardware in critical areas of the plant, to monitor and alarm in the event of unauthorized entry into critical systems.
  —Install a series of passwords or access codes to prevent entry to change program configurations. The system should include a log documenting all entries to the system.
- Installation and maintenance changes
  —Install locks or locking valves in pipelines. The removal of wrenches or wheels from valves to prevent tampering have very limited security value.

—Institute an administrative procedure, such as the management-of-change system or the use of work permits, for jumpering or by-passing critical safety equipment.

3.12.1.2. Equipment Characteristics

A. Electrical and Electronic Devices in Hazardous Areas

NFPA 70 in Sections 500 through 503 defines and classifies the installation of electrical and electronic devices according to an area's potential for fire and explosion. The corresponding NEMA standards for area classification are discussed in Section 3.8.1.

Additional guidance on instrumentation in hazardous areas can be found in ISA RP12.1 and S12.10.

Instrumentation Purging. A hazard classification can be reduced by continuous addition of air or inert gas to a general-purpose enclosure. For example, totally enclosed control panels which contain electronic devices that are not rated for Division I can be used in that service if purged according to ISA S12.4. A purged control panel cannot be opened for adjustments or maintenance if explosive vapors are present or if the components are powered. Electrical power can not be reinstated until the panel has been closed and purged of all explosive vapors.

Intrinsic Safety. Division I areas can also use intrinsically safe barriers to provide an "explosion-proof" installation. These barriers must be located and connected in accordance with the rules for intrinsically safe installations, such as ISA RP12.6.

B. Environmental Effects

Airborne Emissions. ANSI/ISA S71.04 gives a basis for classifying the con-taminants found in emissions. Quantitative data may be difficult to find and to relate to maintenance and performance of instruments. However, there is a set of recognized good practices based on industry experience.

Airborne gases or liquids are primarily a corrosion problem. Airborne solids may present other, more unusual problems, causing magnetic, thermal, electrical, or mechanical failure. For example, an accumulation of dust can insulate an enclosure and cause components to overheat.

Ambient Temperature Effects. All electronic instrumentation has tem-perature limits beyond which it will not operate properly. Ambient tempera-ture for an instrument is the temperature where it is installed; this may be significantly higher than outside air temperature.

High temperatures can result from conduction from nearby heat sources, radiation from the sun or hot equipment, or heat generated by instrument components. These temperatures can be reduced by an air purge, a heat exchanger, or vortex cooling. If cooling is impractical or inadequate, the instrument(s) in question must be moved or respecified and replaced.

Low-temperature limits may be violated by winter conditions or location of an instrument near cold equipment. Some I&C manufacturers supply internal heaters. Use of a heated enclosure is another solution.

ANSI/ISA S71.01 defines environmental conditions for instrument installations. Categories are:

- air conditioned
- enclosed temperature controlled
- sheltered
- outdoor.

An instrument operating within the defined temperature range should achieve the accuracy claimed by the manufacturer. Ranges and zeroes may change slightly with temperature. Most manufacturers can quantify this effect if necessary.

Electromagnetic Interference (EMI). Electrical noise, man-made or natural, can degrade instrument performance and accuracy, even to the point of complete failure. Electromagnetic interference results from magnetic fields. Sources include lightning, electric motors, arc welders, power lines, and radio transmitters. Radios used within a plant are often a source of radio frequency interference (RFI). The solution to EMI is usually a combination of electrical shielding, grounding, signal isolation, and proper location of instrument cables. Points to be considered include:

- requirements for EMI/RFI protection (SAMA standard PMC 33.1 gives a basis for specifying instruments and comparing the degrees of protection offered by different designs.)
- proper grounding of shielded cables, at one point only (preferably near the power source)
- protection of ground wires from traffic, corrosion, and accumulation of debris
- placement of cables far enough from major electrical equipment (e.g., at least five feet)
- placement of signal cables far enough from power cables (e.g., at least a foot)

3.12.1.3. Human Factors

A. Control Room Ergonomics

The designer or specifier of control panels and operator interface systems should consider the human factors (ergonomics) that affect the safe operation of the plant, such as:

- Locating and clustering of related switches, dials, meters, etc. to improve efficiency while at the same time reducing the chance of operator error. ISA RP60.1, RP60.3, and RP60.6 can be helpful in this regard.
- Providing well-organized, uncluttered graphic displays on operator interface stations. ISA S5.3 and S5.5 can provide some guidance on such displays.

B. Operations Interface

A panel board or video display unit (VDU) is often the main source of information for operating personnel, as well as their primary interface with the process. Proper display of information and ease of access to data and controls become vital considerations if a plant is to achieve high efficiency and to operate safely. Poor design and layout promote human error and can create hazardous situations. Some of the types of error that can occur are treated below, together with the precautions that may be taken to combat them.

Observation Error. A panel or VDU allows general observation of a process with specific attention to the important or out-of-specification areas. Errors of observation usually involve reading the wrong item from the display. They also result from operator saturation.

Wrong Indication. This error has the highest probability of occurrence in process control operations. If given two items that have a similar "look-and-feel," the operator may observe one item and mistake it for the other. To reduce the frequency of these errors:

- all devices should be plainly marked with names or tag numbers
- the display can be physically divided into areas, or plainly marked with demarcating lines
- color-coding or colored displays may be of some value (this may be limited by the prevalence of color blindness in operators)
- the display can be enhanced by derived information which combines individual instrument readings
- the display can allow verification of such items as valve positions, operation of motors, etc.

Information Saturation. A display with too much information is confusing. A better practice is to give only the information needed immediately by the operator. Other data can be filed for separate access.

Alarm Saturation. Errors in analysis can result when an operator is asked to cope with too many alarms. Operators often must scan a series of alarms as they occur, deduce what is happening within the process, draw conclusions, and take the appropriate action. Errors of analysis occur when, after properly observing a valid set of data, an operator draws a wrong conclusion or responds inappropriately. These errors can be avoided by assigning priorities to alarms in order to guide the operator.

Lack of Anticipation. This error occurs when an operator finds his process in a satisfactory state but does not foresee a developing problem. If too busy monitoring, say, the first step in a process, the operator may fail to respond to a component failure in a later step.

Manipulation Error. Errors of manipulation occur when an operator draws the proper conclusion from the data but errs in his actions. There have been cases where the right action was taken, but on the wrong control loop. The probability of manipulation error is reduced by good graphic design. The suggestions listed above under "Wrong Indication" also apply here.

3.12.2. I & C Installation Safety

Once the I&C system has been designed, it must be properly installed. Some points to consider are:

- Ensure that the instrumentation and controls have been installed according to the instrumentation and electrical wiring drawings. If not, correct any errors and document any deliberate changes.
- Ensure that the I&C equipment has been installed according to the specifications and vendor installation instructions. If not, document any substitutions.
- Install all control valves with correct orientation.
- Check out all control loops.
- Perform functional checkout and validation.
- Document all changes in the control system configuration.
- Remove all jumpers and bypasses before startup.

3.12.2.1. Loop Checking
Trouble-free startup requires a well planned and disciplined check of field instruments. If hardware is carefully checked against design and installation specifications, later problems can usually be attributed to the control system itself. This allows the startup team to function more efficiently.

One possible approach is a four-step program of checking and testing. The discrete steps are

- visual inspection and verification of installation
- visual and continuity inspection of electrical system
- calibration and operations check
- integrated loop checks

STEP 1—Installation Verification
The first step is visual inspection of each device to verify the following:

- Each instrument is in the proper location and is correctly identified.
- Piping and tubing systems conform to specifications.

- Instrument hookups conform to manufacturers' specifications and installation drawings.
- Control valves and in-line instruments are in the right locations, properly oriented, and spaced correctly for best performance.
- Drains, purges, and bypass and isolation valves are properly installed.
- Orifices and accessories conform to specifications and are properly installed and oriented.
- Instruments and components are accessible for maintenance.
- Instruments and accessories do not interfere with access to equipment.
- Instruments and racks are mounted in accordance with drawings and specifications.

STEP 2—Electrical Verification
The second step is a visual and continuity inspection of each instrument. The following should be checked:

- All conduits, cables, junction boxes, etc., are installed according to applicable specifications.
- Wiring connections agree with loop and electrical drawings.
- Electrical continuity exists between each field instrument and its receiver.
- Instrument wiring is free from grounds and short circuits.
- Alarms and auxiliary components are connected properly.

STEP 3—Instrument Activation and Calibration
Operational checks require electrical power and clean, dry instrument air. This step should establish the following:

- The instruments have been calibrated by the supplier or contractor before installation and tagged accordingly.
- All air supply lines have been blown out. This should be done specifically as part of the precommissioning process, even when lines were tested and blown out during construction.
- All filters contain elements.
- Power supply circuits are energized, and voltages match manufacturers' requirements.
- All process-connected tubing and capillaries have been tested for leaks.
- Instrument scales, charts, etc. have been checked.
- Computer-based equipment, such as the DCS, PLC's, etc., has been checked.

STEP 4—Integrated Loop Check
Thorough loop checking is a complex task that confirms overall system integrity. It can be divided into five phases, performed in the order listed below.

Phase 1—Sequential Loop Check. This assures that each device within a loop is "logically correct." By simulating signals from transmitters to receivers and then to final control elements and all auxiliaries (alarms, interlocks, other components within the loops), the commissioning or maintenance team should verify the following:

- Readout devices correctly follow impressed signals.
- Alarms trip at their set points and proper responses follow.
- Interlocks, shutdown systems, etc., respond as designed and in their proper sequences.
- Control valves respond properly to impressed controller output.
- Controller tuning parameters are set properly.
- On/off valves move in accord with impressed signals.
- Valve positions are accurately fed back to the central system.
- Solenoid valves operate correctly.
- When valves are deenergized, they move to the designated failure positions, and failure alarms are activated.
- Failure alarms are received at the control room, and interlocks and shutdowns operate satisfactorily.
- Motor start/stop controls in the central system send the proper signals into the field.
- Field motor controls function independently.
- Failure alarms, interlocks, and hierarchical systems are properly set.
- Winterizing heaters and tracing systems work properly without interfering with instrument operation or calibration.

Phase 2—Closed Loop Test. This checks the operation of each instrument loop. Proceeding as in the sequential test discussed above, an operator should drive all inputs to the values needed to test each mode of operation of a loop. Most control loops have one manual mode, one hold mode, one or more automatic modes, and one or more fault modes. For each of these, proper operation of all the following must be verified:

- alarms
- displays
- interlocks
- overrides
- data logging
- printouts
- communications with other loops

At the same time, field verification of the operation of output devices, valves, etc. must be made.

Phase 3—Open-Loop Test. This supplements the previous test by checking certain aspects not amenable to closed-loop evaluation.

Hysteresis
- An on/off system should have enough dead band or hysteresis (usually about 1% of set point) to prevent rapid cycling or "chattering" of valves.
- An analog system should not have excessive hysteresis, which deteriorates control action. Some sources of hysteresis are:
  —packing friction
  —undersized actuator
  —defective positioner or linkage

Valve stick/slip
- Control valves should not stick and slip.
- Valves should smoothly track small changes in set point.
- Valves should not move erratically after accumulation of changes or reset wind-up.
- The following possibilities should be checked when a stick/slip problem occurs:
  —undersized spring or actuator
  —packing friction
  —valve binding
  —plugged nozzle in positioner
  —improperly set actuator
  —seal friction in rotary valve

Linearity
- Valve position should accurately follow controller output. Nonlinearity of response can be due to the use of a wrong positioner or incorrect controller tuning.

Signal noise
- All sensor wires should be clean and properly fastened.
- Contacts should be aligned and securely connected.
- Devices and enclosures should be properly supported and anchored.
- Noise filters should be used where necessary.

Phase 4—Tuning. Best control results when the optimum tuning parameters can be determined. This is a complex subject outside the scope of this book. The instrument vendor and publications such as ISA S26 are the best sources of information.

Phase 5—Comparison Test. The previous phases usually reveal the need to make some changes. The user should compare the result of these changes with the original control loop and ask the following:

- Has the loop checking caused changes to be made?
- Are these significant changes that might modify the outcome of previous tests?
- Have tests been repeated where this is the case?

3.12.2.2. Functional Checkout and Validation
With all loops individually functioning, problems still can exist at higher levels of the control system. The tests that remain are therefore on a progressively larger scale.

A. Functional Sequence Validation
This level validates communication among subsystems. The loops and devices should be driven in such a way as to replicate their order of occurrence during production. Once all subsystems are performing in proper sequence, communication is presumed to be validated, and the next level of testing should begin.

B. Administrative Validation
Administrative devices and functions not essential to plant operation include data logging, alarm and event archiving, possible logons and password identifications, system printers, modems, and report generation. The owner/engineer should decide which of these need validation.

C. Integrated System Validation
This is the final test before (or as part of) startup. If possible, it should include a plant-wide simulation (mock batch, "dry run," or water test). This final test should be monitored closely to ensure that all parts of the control system are operating in an acceptable manner.

3.12.2.3. System Configuration Changes
I & C systems frequently are modified after startup. This is especially true with batch processes. Three common types of system change are:

Reparation. This change usually occurs in response to inadequate process performance or a series of production delays. Examples are replacement of improperly sized instruments and reprogramming of control software.

Upgrade. Upgrade is often justified when the original system has become obsolete or "bulky." Whether because of slow speed, fire or electrical hazard, or fear of system failure, a time may come when a device should be upgraded to more modern technology. Proper upgrading at the right time will enhance the safety of operation and prolong the life of a facility.

Expansion. This change includes expansion of the control system itself by automation of a previously manual operation. Reserving space for future systems is appropriate if expansion can be foreseen during design.

3.12.3. I & C Operation Safety

An I & C system will operate safely only if the process runs within the constraints defined by the operating procedures. During normal operation, safety requires the following:

- Maintain parameters within established constraints. Document any variances, intentional or accidental, for investigation. The documentation should describe:
  —deviations from normal flow rate
  —deviations from normal temperature and pressure
  —slow response to changes in operating conditions
  —instability or oscillation at certain conditions
- Periodically check on functionality of instruments and controls, reporting malfunctions to maintenance.

3.12.4. I & C Maintenance Safety

A maintenance program involving both operating and maintenance personnel can help to ensure continued safe operation of an instrumentation and control system. The program should include preventive maintenance supported by systematic monitoring of system behavior. Any observed abnormalities should be reported and investigated as quickly as possible. The maintenance program should provide for:

- periodic tuning of control systems
- replacement of aging control system components
- periodic testing of control valve performance and associated valve maintenance
- replacement of worn or damaged wiring and tubing

The maintenance facility should be consistent with ISA RP52.1 for repair, replacement, and recalibration of instrumentation. The facility should contain:

- spare parts storage consistent with cleanliness requirements for this equipment
- installation and maintenance manual library
- backup software storage

REFERENCES

1. Lewis, G., *Selection of Engineering Materials*, Prentice-Hall, Englewood Cliffs, NJ, 1990.
2. Sandler, H. J., and E. T. Luckiewicz, *Practical Process Engineering, a working approach to plant design*, McGraw-Hill, New York, 1987, Chapters 5 and 17.
3. Uhlig, H. H., and R. W. Revie, *Corrosion and Corrosion Control*, John Wiley & Sons, New York, 1985.
4. Treseder, R. S., *Corrosion Engineer's Reference Book*, National Association of Corrosion Engineers, Houston, 1980.
5. Dillon, C. P., *Corrosion Control in the Chemical Process Industries*, McGraw-Hill, New York, 1986.

6. Port, R. D., and H. M. Herro, *The Nalco Guide to Boiler Failure Analysis*, McGraw-Hill, New York, 1991.
7. Colangelo, V. J., and F. A. Heiser, *Analysis of Metallurgical Failures*, John Wiley & Sons, New York, 1974, Chapter 9.
8. Dillon, C. P., *Material Selection for the Process Industry*, McGraw-Hill, New York, 1992.
9. Nelson, W.E., *Plant Services 14*, No. 4, pp. 17–22 (1993).
10. Pollack, R., in "Fluid Movers: Pumps, Compressors, Fans and Blowers," *Chemical Engineering Magazine*, McGraw-Hill, New York, 1979.
11. Walas, S. M., *Chemical Process Equipment: Selection and Design*, Butterworths, London, 1988.
12. Bloch, H. P., and F. K. Geitner, *Practical Machinery Management for Process Plants, Volume 4—Major Process Equipment Maintenance and Repair*, Gulf Publishing Company, Houston, 1985.
13. Port, R. D., and H. M. Herro, op. cit., p. 205.
14. Ibid., Chapter 14.
15. Dukelow, S. G., *The Control of Boilers*, Instrument Society of America, Research Triangle Park, NC, 1986.
16. Factory Mutual Engineering Corp., *Handbook of Industrial Loss Prevention*, 2d ed., McGraw-Hill, New York, 1967.
17. Lees, F., *Loss Prevention in the Process Industries*, 2 vols., Butterworths, London, 1980.
18. Kletz, T., *Plant Design for Safety: A User-Friendly Approach*, Hemisphere, New York, 1991.

4

GENERAL TOPICS

This chapter covers general plant safety topics not directly related to specific activities or equipment. Principal subjects are the inspection, maintenance, and calibration of equipment (4.1); spare parts handling (4.2); storage and warehousing (4.3); plant modifications (4.4); hazardous operations (4.5); the use of outside contractors (4.6); and the protection of workers by safety gear and the control of the working environment (4.7, 4.8).

4.1. INSPECTION, MAINTENANCE, AND CALIBRATION

In order to ensure safety of the process plant, a comprehensive maintenance program should be in place to certify, inspect, and, where necessary, calibrate equipment and instruments. This program should include:

- *Testing and inspection*—Part of any material and equipment control program is testing or inspection to verify that:
  —materials and equipment as received meet purchase specifications
  —materials and equipment in storage still satisfy requirements
- *Calibration*—Faulty calibration of equipment may allow a plant to operate in a hazardous manner. The control program should therefore include the proper facilities and a schedule for periodic inspection and calibration.
- *Preventive maintenance*—The plant maintenance program should include preventive measures that recognize the limited life of plant equipment and that service or decommission the equipment before it fails and creates a hazard.

4.1.1. Inspection Techniques

Process materials and equipment are inspected in order to reveal defects that would prevent safe operation [1]. Table 4-1 lists various types of defects and their origins; Table 4-2 describes frequently used nondestructive inspection techniques and lists their advantages and disadvantages [2].

TABLE 4-1
Origins and Types of Equipment Defects

- Raw material defects
  —stress cracking
  —gas porosity
  —slag inclusion
  —shrinkage porosity

- Defects produced during manufacture
  —welding defects
  —machining defects
  —heat treating defects
  —residual stress cracking

- Defects produced during assembly
  —additional welding defects
  —missing parts
  —incorrect assembly
  —additional stress cracking

- Defects produced during service
  —wear
  —thermal degradation
  —creep
  —fatigue
  —corrosion

4.1.2. Maintenance Manuals

Vendors will supply instructions for operation and maintenance of each system or piece of equipment. Plant personnel should make sure that the vendor's requirements are practicable and in line with established plant practices. All the relevant information then can be compiled for inclusion in a maintenance manual. The size of the plant and the nature of its organization will determine whether it is better to gather all maintenance information into a central comprehensive set of dedicated maintenance manuals or to file it by equipment item as part of the operating manuals. In either case, the plant must have clearly established lines of responsibility and well understood rules for performing certain types of work under permit (see Section 7.1.4). The maintenance information should cover the following:

- maintenance procedures
- troubleshooting guide
- spare part requirements and specifications
- use of special tools and equipment

TABLE 4-2
Nondestructive Inspection Techniques

Technique	Advantages	Disadvantages
Radiographic: Used to examine the internal soundness of weldments and metals by bombarding the piece with x-rays or gamma rays.	Sharp picture of defects. Film provides a permanent record	Special personnel protection and training required. Both sides of the piece must be accessible.
Ultrasonic: Uses high-frequency sound waves to locate defects.	Very sensitive; can detect very fine surface and subsurface cracks. Equipment is portable Only one side need be accessible.	Personnel must be trained to interpret equipment response. Not effective on rough surfaces or welds with backing rings.
Magnetic particle: Used to find surface defects by applying a liquid suspension of fine particles that flow into fine cracks. A strong magnetic field concentrates the particles in the area of the defect, highlighting its size and shape.	Shows fine cracks that are not noticeable in radiographic examination. Shows where and what material must be removed for weld repair.	Cannot be used on nonmagnetic material. Detects surface cracks only. Cannot detect defects parallel to the magnetic field.
Dye-penetrant: Used for surface defects. A liquid dye is applied to a clean, dry surface and allowed to penetrate surface cracks and dry. A developer put over the surface causes the dye to outline the defect clearly.	Useful for nonmagnetic materials. Can be used on nozzles and surfaces difficult to inspect radiographically.	Detects surface defects only. Not practical on rough surfaces.

- preventive maintenance schedule
  —lubrication
  —cleaning
  —testing
  —calibration
  —replacement
- safety precautions
  —personal protective equipment
  —MSDS for lubricants, chemical, solvents, etc.
  —special procedures

The assigned custodian of the maintenance manuals should see that they are:

- available for all equipment. All relevant information for a particular piece of equipment should be in the appropriate file. This often requires cross-filing.
- kept in a secure location. Copies may be used locally, but a master file should be kept in a designated spot, away from the production area.
- available to all personnel who may be involved in maintenance work.
- up to date. Changes made in the plant should be reflected in the maintenance manuals. Information on new items or changes in procedure should be added, and obsolete information should be purged. These changes should be communicated specifically to maintenance personnel; simply adding them to a maintenance manual is not enough.

4.1.3. Preventive Maintenance

A well-conceived preventive maintenance program is far superior to a reactive approach, because it can prevent many of the breakdowns and plant outages that otherwise would occur and occupy most of the maintenance effort. Work performed on a routine rather than an emergency basis will usually be safer, cheaper, more effective, and of higher quality [3]. The term "preventive maintenance" means different things to different people. It includes, but should go beyond, scheduled lubrication and equipment adjustments. Comprehensive programs include some elements of predictive maintenance. This technique, which may not be justified in small operations, uses evaluation of the actual operating condition of equipment to determine the need for maintenance. Periodic analysis of a lubricating oil, for example, can be used to forecast the time at which it should be changed. Examination of the metallic particles found in the oil can reveal the type of wear which is occurring in machinery and help skilled maintenance technicians to identify a developing problem before it becomes critical. Chapter 3 discusses specific preventive maintenance programs for selected process plant equipment.

4.1.4. Equipment Calibration

If a process plant is to operate safely, the instruments used to control and inspect process systems must be periodically calibrated. Even equipment purchased to a detailed specification may not operate as specified, for the following reasons:

- Equipment received on site may require recalibration because of mishandling during shipment.
- Each item has a definite life cycle which includes a burn-in period, normal operating life, and old age (Figure 4-1). During the burn-in period, the rate of failure is high and the need for calibration is great.

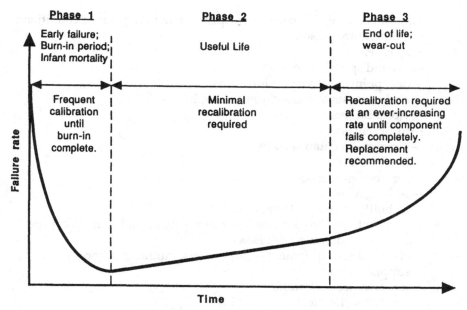

FIGURE 4-1. Component Life-Cycle Curve.

When factory burn-in tests are curtailed, some drift in output still may occur when the equipment is placed in operation in the plant. Equipment must, therefore, be monitored in the early stages of operation to determine whether recalibration is necessary.

- Seasoned process equipment must also be calibrated periodically to compensate for such stresses as:
  —protracted normal operation
  —rapid or greater than anticipated cycling
  —excursions beyond the specified operating ranges
  —physical damage
  —exposure to environmental conditions outside specification

Failure to calibrate may result in failure to detect unsafe operating conditions. Some devices are especially critical to the safety of an operation. These critical devices fall into three categories:

- Equipment designed to warn or to reduce the possibility of unsafe operation
  —combustible gas analyzers
  —toxic gas analyzers
  —oxygen analyzers
  —vibration analyzers
  —radiation monitors

- Equipment designed to maintain operation within specified conditions
  —temperature sensors
  —level sensors
  —pH and specific ion sensors
  —dew point/relative humidity sensors
  —analyzers, such as gas chromatographs
  —vibration sensors
  —RPM sensors
  —pressure/vacuum sensors
  —flow sensors
  —conductivity sensors
  —strain gauges
  —turbidity/smoke detectors
- Equipment used to calibrate items listed above and to inspect piping and process equipment for flaws
  —electrical test equipment (signal tracers, multimeters, power supplies, etc.)
  —temperature calibrators
  —ultrasonic detectors
  —portable gas detectors
  —ground resistance testers
  —pressure/vacuum calibrators
  —X-ray detectors
  —sound level monitors

Table 4-3 lists the calibration procedures recommended by the ISA for specific instruments. The first column identifies the standard or recommended practice which applies in each case.

4.2. SPARE PARTS AND EQUIPMENT

Operating safety is enhanced by the maintenance of a well organized and adequately stocked inventory of equipment, spare parts, and tools. The proper volume of spares is a function of plant size and complexity and of the availability of parts from equipment vendors.

4.2.1. Receiving

Maintaining a quality inventory of spare parts, tools, and equipment requires careful scrutiny of all materials received into stores. Some of the characteristics of a good receiving system are:

- The receiving area has safe unloading facilities for all materials that are expected to be stored, including:

TABLE 4-3	
Instrument Calibration Methods (ISA)	
RP16.6	Methods and Equipment for Calibration of Variable Area Meters (Rotameters).
RP55.1	Hardware Testing of Digital Process Computers.
S12.13, Part I	Performance Requirements, Combustible Gas Detectors.
S12.15, Part I	Performance Requirements for Hydrogen Sulfide Detection Instruments.
S26	Dynamic Response Testing of Process Control Instrumentation.
S37.3	Specifications and Tests for Strain Gauge Pressure Transducers.
S37.5	Specifications and Tests for Strain Gauge Linear Acceleration Transducers.
S37.6	Specifications and Tests for Potentiometric Pressure Transducers.
S37.8	Specifications and Tests for Strain Gauge Force Transducers.
S37.10	Specifications and Tests for Piezoelectric Pressure and Sound-Pressure Transducers.
S37.12	Specifications and Tests for Potentiometric Displacement Transducers.
S75.02	Control Valve Capacity Test Procedure.
S75.13	Methods of Evaluating the Performance of Positioners with Analog Input Signals and Pneumatic Output.
S82.01	Safety Standards for Electrical and Electronic Test, Measuring, Controlling, and Related Equipment—General Requirements—1988.
S82.02	Safety Standards for Electrical and Electronic Test, Measuring, Controlling, and Related Equipment—Electrical and Electronic Test and Measuring Equipment—1988.
S82.03	Safety Standards for Electrical and Electronic Test Measuring, Controlling, and Related Equipment—Electrical and Electronic Process Measurement and Control Equipment—1988.

—sufficient shelter and environmental control

—sufficient and proper material handling equipment

—up-to-date records to allow review of incoming materials and accompanying documentation against purchase orders

- Equipment received from a supplier is compared with the specifications given in the purchase order. If noncompliant equipment is received without satisfactory approval by the originator of the purchase order or other cognizant individual, the equipment may become a safety hazard by its use or installation in a service for which it is not suited. Equipment not in compliance with specifications should be rejected and returned to the manufacturer.

- After receipt, goods are marked and labeled for their intended use and then properly stored. Improper storage can create safety hazards in the following ways:
  —Toxic or flammable materials, such as cleaning solvents, acids, and volatile organics, can expose warehousing personnel to hazards if not properly packaged. This is discussed in Chapter 2 and Section 4.3.
  —Equipment exposed to hostile environments can be damaged and made unsafe to use. A "hostile environment" can range from yard storage without protective covering to not being stored in an environmentally controlled room. Vendors should provide storage instructions to prevent equipment deterioration.
- The receiving process includes a program for the safe and orderly performance of the following activities:
  —unloading from carriers
  —unpacking containers
  —identifying and sorting material
  —checking receipts and purchase orders against packing slips
  —recording receipts
  —noting shortages, damage, defects, etc.
  —maintaining adequate records
  —satisfying quality assurance requirements
  —dispatching material into stores or to the use area
  —labelling and marking parts for intended use

4.2.2. Storage

The storage area should have the following features:

- items in stores kept in safe reach
- sufficient space allocated for tools and spare parts
- parts with different materials of construction stored in separate locations to prevent confusion and mistaken installation of an item in the wrong location or service (such mix-ups can also be prevented by color coding the parts, tags, or storage facilities)
- ample aisle space
- shelf space adaptable to changes in sizes and quantities of stored items

Other desirable characteristics are:

- handling equipment large enough and strong enough for all items to be stored
- stored material protected from deterioration by chemical exposure, weather conditions, or excessive heat, cold, or humidity
- simple and efficient identification and retrieval of stored material
- oldest stock used first

4.2.3. Disbursement

Downtime not only reduces production but also can create hazards. Since maintenance work can not always be planned, some minimum inventory of spare parts should be available for disbursement and routing to the trouble spot in the plant. For effective disbursement:

- Lists of all materials, tools, and equipment necessary for each specific task should be prepared and submitted to stores.
- Stores should be able to identify, locate, and disburse these materials quickly.
- Stores should be aware of any special handling requirements, identify them to maintenance personnel, and provide handling equipment as required.
- Stores should keep up-to-date records of all disbursements and should actively pursue the recovery of tools, equipment, and unused materials.

All returns should be inspected for damage and defects before replacement in stores. Damaged items should be repaired or scrapped.

4.3. STORAGE AND WAREHOUSING

This section is concerned with safety in central storage areas such as warehouses. It covers the movement of containers and materials into and out of storage, safe practices during storage, and the operation of vehicles such as lift trucks and the tractor-trailers used to ship raw materials and products. Comprehensive fire protection is especially important in warehouses [4] and requires proper design of automatic sprinkler systems, adequate water supply, availability of large and small hoses as well as hydrants, location of proper portable fire extinguishers, procedures for fire fighting, a well-organized plant emergency organization, and a working fire prevention program. Specifics are covered in various NFPA codes [5] and other standard references [6].

There are a several general actions which help to protect facilities from disasters:

- Employees must be aware of the fire hazards of the commodities that they are handling and of the problem of storage facility fires. Communication must flow freely up and down internal channels for personnel awareness.
- Fire protection engineers can evaluate hazards and protective systems and advise with regard to the technical issues of loss prevention.
- Keep the fire department and corporate management apprised of changing commodities and storage configurations which might reduce the ability of the sprinkler system to control a fire. Anticipate the effects of change.

- Act upon insurance carrier recommendations.
- Prepare pre-incident information and plans for emergencies, working closely with the local fire department.
- Learn the fire-protection implications of:
    —Storage configuration changes.
    —Commodity classes, which types of storage are most hazardous, and how dangers can be controlled.
    —Fire areas and size of open areas related to fire spreading.
    —Fire walls.
    —Building construction and its relation to possible early collapse.
    —Material handling practices and possible causes of fire ignition.
    —Manual (internal) fire fighting, with adequate and realistic training.
    —Environmental impact, with water runoff possibly contaminating the public water supply, forcing the withholding of water from the fire department.

4.3.1. General Storage Techniques

4.3.1.1. Protection from Ambient Hazards

The safest storage facility is a completely protected single-story building, well separated from other buildings. The only ignition sources to which materials stored in such a facility would be exposed are those from material-handling equipment.

Within a given building, storage and manufacturing operations should be separated by fire compartments or cutoffs to prevent spread of fires from one area to another. A fire compartment is part of a building which is separated from the rest by fire-resistant walls, ceilings, and fire stops such that a fire can be contained and not spread to adjacent rooms, other floors, or any other part of the building.

In multistory buildings, upper floors are preferred for storage, because firefighting water could leak from a process-area fire on a higher floor. Materials which must be stored below potential fire sites should be protected by waterproof tarpaulins.

The NFPA addresses general indoor storage in Standard 231.

Outdoors, material should be stored under cover if protection is needed from wind, rain or flooding, heat, and ignition from the sun or nearby sparks and fires, or vandalism and theft. Special precautions are necessary with some commodities. These include carbonaceous materials (coal, coke, charcoal), cellulose and fibers (roll paper, pulpwood, wood chips, wastepaper, baled fibers), and rubber (tires, bales). Refer to the NFPA guides for these types of materials.

4.3.1.2. Stacking Procedures

Major warehouse fires have occurred even in facilities with automatic sprinkler systems and other built-in fire protection. Problems often are due to lack of awareness of hazards and human behavior which compromises the effectiveness of the fire-protection systems. Whenever the conditions of storage change, those involved in handling materials or managing the storage facility must review the impact of the change on the adequacy of the protective systems. Such changes include increases in the quantity or height of storage and changes in the type or arrangement of materials stored. Some frequently used methods of storage are discussed below.

Bulk Storage

Unpackaged materials in loose, free-flowing condition (e.g., powder, granules, pellets, or flakes) which may be stored in silos, bins, tanks, or piles. These materials may be moved by belt conveyors, air-fluidized transport, or bucket conveyors, any of which can cause hazardous dust formation.

Solid Piling

Cartons, boxes, bales, and bags can be piled up in direct contact. Bagged material should be cross-tied, with the mouths of bags toward the inside of the pile. Bags should be stepped back one row if piled above 5 feet high and stepped again for each additional 3 feet. Sacks should never be removed from lower layers in a way that could undermine the stability of the pile. Air spaces or flues may be left between packages or piles.

Solid piling is comparatively fire resistant, but flammable outer surfaces can be hazardous. Piles more than 15 feet high limit access and should be avoided. Solid piling of general goods up to 12 feet requires sprinkler protection per NFPA Standard 13. NFPA Standard 231 defines protection required for plastics over 5 feet high and other stacks up to 30 feet high.

Drums and barrels should be stored in racks. When they are stacked on end, preferably in the shape of a pyramid, there should be separators between rows. If drums are stored on their sides, piles should be symmetrical and stable, also pyramidal, with the bottom row blocked securely to prevent rolling.

Palletized Storage

Unit loads mounted on pallets are very common, usually 3 to 4 feet on each dimension. Most pallets are wood, but metal, solid or expanded plastic, and paperboard are also available. Pallets are usually moved by pallet jack or forklift truck. Permissible stacking height depends on the crushing resistance of the palletized material, the building configuration, and fire protection considerations. Air spaces formed by the pallets can themselves help to spread fires. Sprinkler system requirements are found in NFPA Standards 13 and 231.

Empty pallets can be a considerable fire hazard, a fact that is often ignored. The stacking height of wood or nonexpanded polyethylene solid deck pallets should be limited to 6 feet, and they should be kept in a small area in detached piles. Other types of plastic pallets should be piled no more than 4 feet high, with no more than two stacks together. They should be kept 8 feet from other pallets and 25 feet from other commodities. Idle pallets should not be stored in racks and are best stored outdoors at a safe distance from building walls.

Rack Storage

Storage racks are structural frameworks which support unit loads, usually on pallets. Storage racks can be built very high, say to 100 feet, with automatic or manually operated machinery to access the storage locations. There are many variations in height and configuration, with single- and double-row racks. Large numbers of air flues between unit loads in racks, empty spaces, partial loads, and narrow aisles can allow fires to spread quickly. Heat distortion of precision stacking machinery and structure can impair the entire warehouse operation. Extensive in-rack sprinkler coverage is therefore necessary to protect the installation.

Materials must be stacked at safe heights, with adequate access for normal movement and emergencies. Figures 4-2, 4-3, and 4-4 show some of the key parameters for stacking, accessing, and arranging materials in storage. NFPA requirements for rack storage are in Standard 231C.

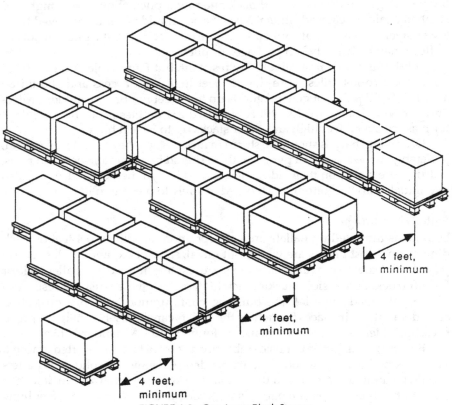

FIGURE 4-2. One-Layer Block Storage.

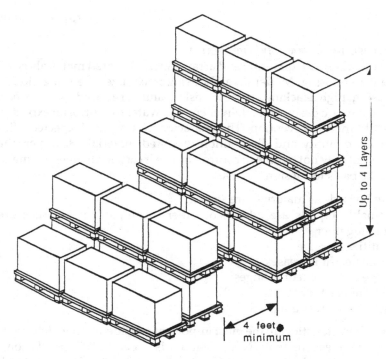

FIGURE 4-3. Block Storage. Four layers or up to 16 feet in height; minimum aisle width 4 feet.

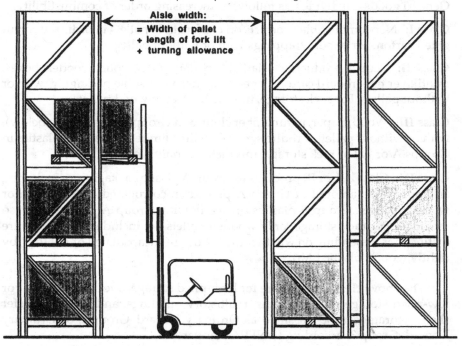

FIGURE 4-4. Rack Storage Using a Conventional Fork Lift Truck.

4.3.1.3. Storage Techniques and Fire Protection

Consideration of content, quantity, and arrangement of stored materials is the key to safe warehousing. Modern warehousing maximizes the utilization of space by using high stacking, narrow aisles, and large undivided areas. Sprinkler systems are designed to deliver enough water to control an expected fire without human intervention. The characteristics of the "expected" fire depend on the quantity and distribution of stored material. However, the sprinkler system may not be able to control the fire if certain changes are made away from the basis for design, such as:

- more-hazardous materials are stored
- flammable liquids are added, or other incompatible materials are stored together
- quantities increase
- storage height increases
- storage arrangement changes
- aisles are used for storage
- solid rack shelves are used

Material characteristics are the most important consideration in determining fire-protection requirements. NFPA Standards 231 and 231C classify commodities by heat of combustion, rate of heat release, and rate of flame spread. General goods are divided as follows in increasing order of combustibility:

Class I: Noncombustible products on combustible pallets in ordinary corrugated cartons, or paper wrappings with or without pallets.

Class II: Class I products in slatted wooden crates, solid wooden boxes, multilayer paperboard cartons, or equivalent combustible packaging, with or without pallets. For rack storage, this includes only wood pallets.

Class III: Wood, paper, natural fiber cloth, and Group C plastics (see below), with or without pallets. Products may contain a limited amount of plastics in Groups A or B. For rack storage, this includes only wood pallets.

Class IV: Class I, II, or III products in Group A plastic packaging, or including an appreciable amount of Group A plastics in corrugated cartons, with or without pallets. Group B plastics or free-flowing Group A plastics are also included. For rack storage, only wooden pallets are included, and there are distinctions depending on whether the goods are encapsulated or not (e.g., by tightly wound plastic film).

The above classifications are for high-piled storage, over 12 feet high. For low-piled storage in stockrooms, multistory buildings, and freight transfer sheds, storage can be classified as Ordinary Hazard–Group 2 or Ordinary Hazard–Group 3, according to NFPA Standard 13.

Plastics and Elastomers
Most synthetic polymers generate about twice as much heat per unit weight as wood, paper, or cotton cloth. In the four NFPA commodity classes, polymers are divided into three groups according to their burning rates. Unless noted otherwise, the list below refers to unmodified polymers. Burning rates can be influenced by physical form and the use of flame retardants.

Group A: ABS, acetal, acrylic, butyl rubber, EPDM, FRP, expanded natural rubber, nitrile rubber, PET, polybutadiene, polycarbonate, polyethylene, polypropylene, polystyrene, polyurethane, highly plasticized PVC, SAN, SBR.

Group B: Cellulosics, chloroprene rubber, certain fluoroplastics (ECTFE, ETFE, FEP), unexpanded natural rubber, nylon, silicone rubber.

Group C: Fluoroplastics (PCTFE, PTFE, PVF, PVDF), melamine, phenolic, lightly plasticized PVC, PVDC, urea-based resins.

Besides material characteristics, other important considerations are:

- storage height
- aisle width
- segregation of hazardous materials
- system design variables

Storage Height
Storage height is a critical variable which affects the spreading rate and difficulty of control of a fire. In tests, fire intensity varied with the square of storage height. For example, a 10% increase in rack storage height from 20 to 22 feet requires a 35% increase in sprinkler discharge density. All heights up to 15 feet in solid piles or 12 feet in palletized and rack storage require the same level of sprinkler discharge, except for plastics. In-rack sprinklers for rack storage avoid the need for height limitations. The maximum height of storage should be 18 inches below the sprinkler deflectors.

Aisle Width
Wide aisles help distribution of water, minimize "jumping" of a fire, and allow space for firefighting and salvage. It is desirable for aisle width to be half the pile height. Palletized storage piles should be less than 50 feet wide between major aisles, so that hose streams can reach centers of the piles. Aisles 4 to 8 feet wide are common for racks; widths less than 4 feet require more protection.

Segregation
Automatic fire protection systems can be overwhelmed by the very rapid spread of fires in flammable or explosive materials. Wood, cardboard, paper, and plastic packaging can also cause the rapid spread of fire. The proper segregation and isolation of such materials is essential in preventing catastrophes.

The NFPA and insurance firms recommend that flammable liquids and aerosol containers be separated from general storage. Hazardous materials stored in chemical operations must be identified and dealt with appropriately.

Design Variables
These include:
- use of in-rack sprinklers
- temperature rating of ceiling sprinklers
- clearance below sprinklers
- dry pipe or wet pipe sprinkler systems
- pile stability
- packaging and encapsulation
- smoke venting
- use of high expansion foam
- sprinkler orifice size
- sprinkler response time

4.3.1.4. Storage Categories
Warehoused goods can be classified into several categories. Table 4-4 lists these, generally in the order of most to least hazardous. The table gives examples of each category and general recommendations for their storage.

4.3.2. Stored Materials and Containers

Preventing exposure of personnel to toxic substances is of primary concern. These substances can be gaseous, liquid, or solid. They can enter the unprotected body by inhalation, skin absorption, ingestion, or contact with a wound. Effects can be at the point of contact or systemic, and they can be acute or chronic. Acute exposures result in symptoms immediately or shortly after exposure. Chronic exposures can produce long-term effects, which vary with the substance, its concentration, and the number and duration of exposures.

Toxic effects may be temporary and reversible, permanent and disabling, or lethal. Exposures can produce obvious symptoms (e.g., burning, coughing, nausea, rashes), but even without such symptoms, an exposure can produce toxic effects.

Dangerous conditions can be caused by:

- explosion and fire
  —chemical reaction
  —ignition of explosive or flammable materials
  —ignition of materials because of oxygen enrichment
  —agitation of shock or friction-sensitive materials
  —sudden release of materials under pressure

TABLE 4-4
Categories of Warehoused Goods

Hazard Category: Characteristics of Material Stored	Material Examples	Recommended Storage Location	Quantities and Limits
Explosive; very highly flammable substances	picric acid, blasting materials	Single level structure, in separate fire compartments, pressure relief laterally or upward	max. 50 tons per fire compartment; max. 300 tons per building; block storage max. 2 rows or 8 ft deep, one layer or 4 ft high; min. 4 ft between blocks
Autoreactive; spontaneously flammable substances	ethylene oxide, acrylic esters, phosphorus, catalysts	Single level structure, in separate fire compartments	max. 50 tons per fire compartment; max. 300 tons per building; without storage racks max. 2 rows or 8 ft deep, one layer or 4 ft high; min. 4 ft between blocks
Water reactive substances, causing ignition or production of flammable gases	alkali metals, metal hydrides, metal amides, metal oxides, aluminum powder, magnesium powder, zinc dust	Preferably single level structure; in separate fire compartments; no contact with water permitted via piping or water connections	max. 100 tons per fire compartment; max. 500 tons per building; block storage max. 2 rows or 8 ft deep, max. 3 layers or 12 ft high
Water reactive substances, causing production of toxic gases	acid chlorides, thionyl chloride, aluminum chloride, oleum	Preferably single level structure; in separate fire compartments; no contact with water permitted via piping or water connections	max. 250 tons per fire compartment; max. 500 tons per building; block storage with pallets max. 2 rows or 8 ft deep, max. 3 layers or 12 ft high
Combustion supporting substances; oxidizing agents; peroxides	concentrated hydrogen peroxide, sodium nitrate, ammonium nitrate	Separate fire compartments; can be stored with noncombustible substances up to total permissible level	max. 250 tons per fire compartment; block storage with and without pallets max. 2 rows or 8 ft deep, max. 4 layers or 16 ft high; min. 4 ft between blocks
Flammable liquids with flash points 130°F or less in mobile containers, including tank cars up to 1300 gal	organic solvents such as acetone, toluene, and aqueous solutions with over 6% ethanol	Separate fire compartments; can be stored with low hazard substances up to total permissible level	**With automatic extinguishing system:** max. 250 tons per fire compartment; block storage with and without pallets max. 2 rows or 8 ft deep, max. 4 layers or 16 ft high; min. 4 ft between blocks; **Without automatic ext. system:** max. 10 tons per fire compartment; block storage with and without pallets max. 2 rows or 8 ft deep, max. 2 layers or 8 ft high; min. 4 ft between blocks

TABLE 4-4 *(Continued)*
Categories of Warehoused Goods

Hazard Category: Characteristics of Material Stored	Material Examples	Recommended Storage Location	Quantities and Limits
Combustible solids	Prussian blue, sulfur, ammonium dichromate, active charcoal	Separate fire compartments; can be stored with low hazard substances up to total permissible level	max. 250 tons per fire compartment; block storage with and without pallets max. 2 rows or 8 ft deep, max. 4 layers or 16 ft high; min. 4 ft between blocks
Toxic substances; substances which generate very toxic or strongly malodorous fumes in fire; water polluting substances; highly malodorous substances	polychloro-dibenzofurans, dioxins, hydrochloric and sulfuric acids above 10% conc.	Separate fire compartments; can be stored with low hazard substances up to total permissible level	max. 250 tons per fire compartment; block storage with and without pallets max. 2 rows or 8 ft deep, max. 4 layers or 16 ft high; min. 4 ft between blocks
Compressed and liquefied gases: (A) mobile pressure vessels (B) spray cans and aerosol dispensers	ammonia, chlorine, crop-protection sprays	(A) Separate fire compartments or in open air protected from the weather (B) Separate fire compartments; can be stored with low hazard substances up to permissible level Especially toxic gases such as ethylene oxide, chlorine, and phosgene must be kept under lock	size and number of containers per pressure vessel regulations; (A) toxic gases in "rolling tanks" max. 5 tons per store or per building (B) spray cans and aerosol dispensers max. 250 tons per fire compartment; block storage with pallets max. 2 rows or 8 ft deep, max. 3 layers or 12 ft high; min. 4 ft between blocks
Low hazard substances: (A) combustible solids; substances in combustibility classes 2&3, empty containers, pallets (B) combustible liquids; flash point more than 55°C (C) noncom- bustible solids; substances in combustibility class 1, empty containers (D) noncom- bustible liquids; no flash-point (e.g. water)	alkali hydroxides, calcium oxide, common salt, active charcoal (class 2 or 3)	Can be stored together with some of above substances, as described above	Block storage max. 4 layers or 16 ft high; min. 4 ft between blocks (A), (B): max. 250 tons per fire compartment w/o fire prevention, max. 800 tons w/ fire prevention; block storage max. 2 rows or 8 ft deep, (C), (D): no limits when stored alone; block storage max. 4 rows or 16 ft deep

- oxygen deficiency
  —displacement of oxygen by another gas
  —consumption of oxygen by a chemical reaction
- ionizing radiation
- biologic hazards
  —disease-causing organisms from research or production facilities
  —release of genetically active material
- physical hazards
  —improperly positioned or supported objects that might fall
    (e.g., drums, boards)
  —sharp objects (e.g., nails, metal shards, broken glass)
  —slippery surfaces
  —uneven or damaged floors
  —steep ramps
  —heavy equipment operations
  —transport operations
  —damaged protective equipment
  —manufacturing processes (e.g., mixing of chemicals)
  —electrical hazards (e.g., power lines, wires, cables)
  —heat stress for workers wearing protective clothing
  —noise causing distraction, hearing loss, or impaired communication

Good storage practice for hazardous materials includes segregation or separation from other materials. Separate detached structures, isolated fire compartments, and separation by inert or nonhazardous materials are recommended, depending on the degree of hazard. Other points to check are the need for pressure relief vents for containers and explosion venting for the storage area, which should be a separate structure.

The handling and storage of explosive materials are stringently controlled and regulated. Fireproof, bulletproof, separated, and secured magazines are used for storage. They must be clean, dry, and well ventilated, with openings less than 110 in.^{2} that are screened to keep out sparks and animals. No sources of ignition, including matches, flammable materials, metal tools, or other metal (such as exposed nail heads) are permitted. No electricity is permitted in a magazine other than artificial light from an approved battery-operated lantern or flashlight. Explosives should not be exposed to direct sunlight. No fire or sparks are permitted near the building, and the surrounding area should be cleared of any grass or flammable substances. When retrieving materials, the oldest explosives should be used first. Packages should be opened at least 50 feet away from the magazine, and then only with nonsparking tools (e.g., wood wedges and wood, fiber, rawhide, zinc, babbitt or rubber mallets).

Only authorized workers and those operating under permit should enter the area. Safety and emergency facilities should include barriers, traffic control, blast protection, fire control, neutralizing materials, and protective clothing.

4.3.2.1. Labeling Requirements

All storage areas should conspicuously display material identification signs. Hazardous materials should be identified according to NFPA Standards 704, 49, and 325M. When several materials are stored in a common area, hazard identification should be for the most serious in terms of health, flammability, and reactivity. NFPA 704 defines the use of a diamond pattern (see Figure 2-3), which identifies the nature of a hazard and allows proper protective equipment to be used. Specific concern should be given to chemicals which are oxidizing, combustible, unstable, water or air reactive, corrosive, self-heating, or toxic.

Packages should be marked specifically with the name of the product and hazard information. Other information normally includes name and address of supplier, net weight, and process information such as batch, blend, or lot number, date of manufacture, and other "variable" information. Shelf life information, where applicable, is also appropriate. International shipments should be labeled in accordance with regulations at the destination and with any specific agreements made with the customer. Labeling of air shipments should conform to any additional requirements set by the International Air Transportation Association (IATA).

Material Safety Data Sheets (MSDS) must be available for all materials, and personnel who handle the hazardous materials should be trained regarding dangers, handling, and accident and cleanup procedures.

4.3.2.2. Protection from Internal Contamination

Containers should be kept sealed during storage. Water vapor, dust, and other contamination must be excluded, and tearing or piercing of containers avoided. Spilled material must be rapidly cleaned up and disposed of or resealed.

4.3.2.3. Shelf-Life Concerns

Materials which can deteriorate, become unstable, lose potency, or degrade into hazardous compounds must be clearly marked. The responsible manager must keep track of these materials and take appropriate action to dispose of or replace them. Nonhazardous deterioration of materials can be dealt with in a conventional manner. The formation of hazardous materials demands extra attention and due care.

Disposal of "expired" or out-of-date materials must be in accord with regulations for the type, quantity, and hazard level of the material. Information regarding safe and proper disposal should be obtained before it is needed, not afterward.

Unstable chemicals subject to decomposition or other dangerous reactions (e.g., organic peroxides) require special precautions. The plans for handling such materials should also consider:

- catalytic effects of containers
- materials in the same storage area that could react with each other
- presence of inhibitors
- effect of direct sunlight or temperature changes

4.3.2.4. Temperature Extremes in Storage
Thermally sensitive materials must be stored in such a way that they will not react or become dangerous. Certain materials must be kept under refrigeration to maintain stability. Extreme ambient temperatures or direct exposure to sunlight can affect some materials.

If storage area temperature moves toward an extreme for any reason, steps must be taken to locate and remove problem materials as quickly as possible. In case of fire or other emergency, safe measures should follow a previously prepared plan.

4.3.2.5. Testing for Inhibitor Activity
Small quantities of chemical inhibitors added to fuels, oxidants, or combustible solids can retard ignition, flames, and spreading capabilities. Plastics and other compound materials also use inhibitors to retard undesired chemical reactions.

If chemical safety and stability depend on an inhibitor, it is important that its effectiveness be maintained. Should the inhibitor deteriorate with time, knowledge of any changes in its characteristics should be obtained by testing.

4.3.2.6. Damaged Containers
Damaged or open containers pose hazards and must be found and fixed as soon as possible. Finding such containers is the key, and regular inspections of storage areas will maximize the probability of timely discovery. Materials moving to or from storage should be inspected at the time of each movement. Damage or loosening of covers or seals can result from jostling, bumping, falling, striking objects, or other motion during material handling.

Key factors to be identified in dealing with an incident are:

- nature of the problem
- conditions which modify the size or effects of a spill
- potential losses
- possible control measures

When these are identified, objectives and control tactics can be chosen. This process has been defined as a Human–Machine System by the NFPA [7] and is illustrated in Figure 4-5.

The type, condition, and behavior of the damaged container must be determined. If not already failed, it may be weakened from thermal, mechanical, or internal chemical stress. Prevention of container failure is the primary objective. Steps to prevent failure include container cooling, barrier placement, and removal of adjacent materials. Danger can be controlled by:

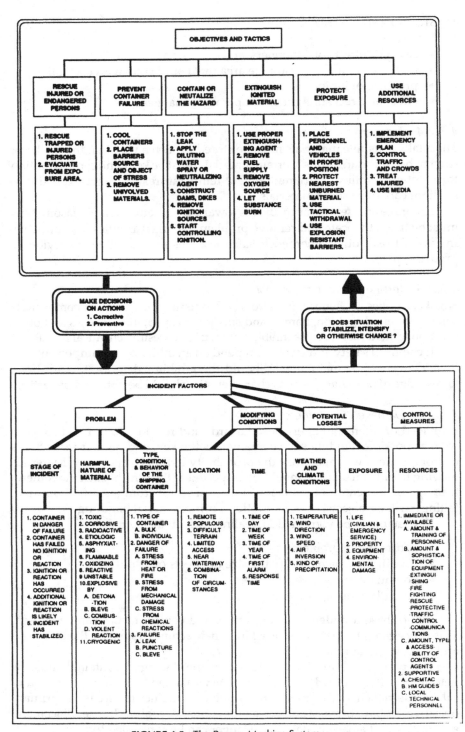

FIGURE 4.5. The Person–Machine System.

- stopping a leak
- spraying with water
- damming, diking or channeling
- removing ignition sources
- controlled ignition

4.3.2.7. Preshipment Inspection of Containers

Containers of hazardous material are subject to strict regulations on labeling, packaging, and mode of transport, as described in earlier sections. To assure compliance, containers should be inspected prior to shipment for the following:

- material identified and classified as to nature and degree of hazard
- containers labeled correctly, including as appropriate DOT label, name of product, hazard information, name and address of supplier, net weight, process information (batch, blend, lot number), date of manufacture, shelf life, etc.
- hazard information sufficient for carrier to properly placard the vehicle and supply (as appropriate) the 704M diamond identification (see Figure 2-3).
- containers constructed to specifications and design requirements
- containers sound and fully sealed
- containers supported, cushioned, or secured where necessary to prevent damage in transit
- shipment accompanied by shipping papers which list names and quantities of all hazardous materials, with appropriate hazard and emergency handling information

4.3.2.8. Technical Assistance

Technical assistance may be advisable or required for hazardous materials emergencies. DOT coordinates all movement of hazardous substances its agencies include:

- Materials Transportation Bureau (MTB), which coordinates implementation of regulations listing hazardous materials and controlling their labeling, packaging, and transport by highway, rail, aircraft, or vessel (49CFR171-179)
- Federal Highway Administration (FHWA) Bureau of Motor Carrier Safety (BMCS), which inspects facilities of shippers as well as trucks on the road
- Federal Railroad Administration (FRA), which inspects hazardous materials shipped by rail as well as railroad equipment and infrastructure
- U.S. Coast Guard (USCG), which inspects hazardous shipments by water and provides:
  —National Strike Force, trained to contain, clean, and dispose of hazardous material spills on the water

—Chemical Hazards Response Information System (CHRIS), provid ing hazardous material information in a four-volume manual, the Haz ard Assessment Computer System (HACS), and the National Response Center (NRC), a regional response team activator and information source

Other incidents could involve agencies such as:

- National Transportation Safety Board (NTSB), which investigates and studies accidents.
- Environmental Protection Agency (EPA), which is responsible for water and air quality, solid waste, and pesticides; the FWPCA regulates spills of oil and hazardous substances and requires immediate reporting. Includes Regional Response Team and Environmental Response Team assistance, and a computerized database, OHMTADS (Oil and Hazardous Materials Technical Assistance Data System).
- Department of Energy (DOE), Federal Radiological Monitoring and Assistance Plan (FRMAP) for emergencies involving radioactive materials.
- Nuclear Regulatory Commission (NRC), which regulates use of radioactive materials; permits DOT to regulate transport and jointly established regulations for packaging, labeling, handling, loading, and storing radioactive materials during transportation.
- State government agencies.
- Local government agencies.
- Other organizations:
  —Chemical Transportation Emergency Center (CHEMTREC), established by the Chemical Manufacturers Association (CMA) to provide information and assistance at any time.
  —Chlorine Emergency Plan (CHLOREP), established by The Chlorine Institute to provide information and assistance for chlorine incidents.
  —Pesticide Safety Team Network (PSTN), coordinated by the National Agricultural Chemicals Association to help with incidents involving Class B poison pesticides.

4.3.3. Material Movement

4.3.3.1. Proper Handling Techniques
Hazards can arise from improperly designed tools or work areas, improper lifting or reaching, poor visual conditions, and repetitive motion in an awkward position. Excessive fatigue and discomfort will lead to pain and soreness and eventually to accidents. Protective requirements have been set in many OSHA and NIOSH Standards.

Improper lifting often causes injuries such as pulled muscles, disk lesions, and hernias. Safe lifting generally requires the body to be configured in the following six ways:

- feet parted, with one alongside and one behind the object
- back straight, as nearly vertical as possible
- arms and elbows tucked in, bringing object close
- entire hand gripping the object
- chin tucked in
- body weight kept directly over feet

All lifting, pushing, pulling, and carrying must be done with care. For example, one significant cause of injury is a sudden twisting motion while lifting or carrying.

Static or isometric work done in a cramped posture can fatigue muscles. Support for rigid body positions should be supplied (e.g., armrests).

Personal protective equipment may be necessary when handling hazardous materials, as described in Section 4.7.1. PPE is a second line of defense, and it requires extra care when working with such devices.

The dangers of the material handling process itself can be magnified by faulty techniques or poor layout. When using material-handling equipment, the operator must be aware of building characteristics such as ceiling heights, aisle widths, overhead obstructions, floor load capacities, doorway clearances, positions of columns, and other structural limitations.

Equipment and facility design or modification can make operations safer. Some examples are:

- counterweighted conveyor gates
- curve guards on conveyor turns
- nonskid treads between rollers on conveyors for crossing or inspection
- fork lift truck crush bar protection
- dock levelers and truck anchors
- easy and safe flow of materials between areas
- adequate aisles for simultaneous pedestrian and vehicular traffic
- adequate space for repair and maintenance
- conveyor height requiring minimal stretching to place or remove items

If these considerations have not been designed into a plant, much greater precautions should be taken by operating personnel.

Personnel must be trained in a wide variety of safe practices. In the field of material movement, these would include procedures for:

- handling materials
- unloading materials from drums
- use of tools
- operation of vehicles
- use of freight elevators
- selection and use of protective gear

4.3.3.2. Pallet Specification and Condition

Pallets come in many sizes. Because of shipments in trailers and boxcars, a standard size of 42 × 48 inches is most common. Traditional pallets are of hardwood joined by nails and adhesives. Other materials such as plastics and various composites also are used. Treated pallets and those constructed of materials which retard burning are safer in the storage environment.

Damaged pallets may be unsafe because of sharp projections or broken parts; these pallets should be discarded in a safe manner.

Stability of their loads should be verified whenever pallets are received or moved to storage. Interlocked packages and wrapped or strapped loads are most secure against shifting, falling, or other movement during pallet handling (Figure 4-6).

4.3.3.3. Container Handling

Solids and liquids are packaged and handled in bags, fiber drums, steel drums, semi-bulk containers, and bulk containers. The common 50- and 100-lb paper or cloth bags must be handled carefully to avoid mechanical damage. Slip-over bags should be kept available to prevent material loss and dust release from broken bags.

Bags are often opened with knives. These should have hilts or guards and should be kept in scabbards when not in use. Boxes and cartons must be opened with care to avoid wire punctures, nail punctures, or cuts. Banded or

Block Pattern

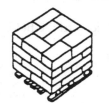     

Brick Pattern   **Pinwheel Pattern**

FIGURE 4-6. Pallet Stacking Patterns.

strapped containers should never be handled by the bands. The bands are under tension and frequently sharp, and they should be broken only with proper tools and the correct procedures.

Containers should be placed on pallets or platforms to prevent water damage and protected from the weather. Bags on pallets should be interlocked to avoid instability when moved.

Drums and barrels, whether lined or unlined, should be filled and emptied carefully to minimize dust. Flammable, toxic, and corrosive materials and those with very fine particle size must be handled with extra care and appropriate safeguards. Sometimes the material can be moistened to suppress dust. For others, hoods, exhaust ventilation, or even complete enclosures must be provided.

Reactors and other vessels can be charged either directly from containers of raw materials, or with pre-weighed charges inserted by special charging devices or hoppers. The appropriate loading technique depends on the hazardous nature of the charged material.

4.3.3.4. Powered Industrial Trucks

Powered industrial trucks come in many shapes and sizes, with a variety of power sources (battery/electric, gasoline, LPG, diesel fuel, etc.) and load engagement options. They may be operated by robots, riders, or persons walking alongside.

Careless or uninformed operation of powered trucks can cause major damage. Common accidents include collisions with sprinkler piping, fire doors, and fire protection equipment; toppling of loads stacked to excessive heights; and careless handling of hazardous materials. It is very important to have adequate operator training, regular and thorough truck maintenance, clear passageways for truck traffic, clearly marked and protected (if feasible) overhead and exposed piping.

Standard trucks are fully equipped with proper guards and safety devices (overhead protection or enclosures, guarded drives and tires, distinctive paint color, warning horns, truck-mounted fire extinguishers, etc.). Trucks used in hazardous locations must be specifically approved for their service. Further, each type of truck needs special handling for recharging or refueling. Table 4-5 summarizes recommendations for various applications, and NFPA Standard 505 contains further information.

Safe operation of powered trucks requires observance of a series of guidelines:

- Operate at a safe speed for the existing conditions, allowing stopping within a certain clear distance ahead and turning without chance of overturning.
- Establish speed limits, often 6 mph for loaded vehicles, and use governors where practical.

TABLE 4-5
Recommended Trucks for Various Occupancies
(Source: Factory Mutual System)

Location	Typical Occupancies	Types of Trucks[a] Approved and Listed[b]
Indoor or outdoor locations containing materials of ordinary fire hazard	Grocery warehouse Cloth storage Paper manufacturing and working Textile processes except opening, blending, bale storage, and Class III locations Bakery Leather Tanning Foundries and forge shops Sheet-metal working Machine-tool occupancies	Electrical—Type E Gasoline—Type G Diesel—Type D LP-Gas—Type LP Dual Fuel—Type G-LP
Class I, Division 1[c] Locations in which explosive concentrations of flammable gases or vapors may exist under normal operating conditions or where accidental release of hazardous concentrations of such materials may occur simultaneously with failure of electrical equipment	Few areas in this division in which trucks would be used	Electrical—Type EX (Class I, Division 1, Group D only; no truck should be used in Groups A, B, and C.) Gasoline, diesel, and LP gas not recommended for this service
Class I, Division 2[d] Locations in which flammable liquids or gases are handled in closed systems or containers from which they can escape only by accident, or locations in which hazardous concentrations are normally prevented by positive mechanical ventilation	Paint mixing, spraying, or dipping Storage of flammable gases in cylinders Storage of flammable liquids in drums or cans Solvent recovery Chemical processes using flammable solvents in closed equipment Rubber-cement mixing	Electrical—Types EE, EX[e] Diesel—Type DY[e] Gasoline, diesel Types D & DS, and LP gas not recommended for this service
Class II, Division 1[c] Locations in which explosive mixtures of combustible dusts may be present in the air under normal operating conditions, or where mechanical failure of equipment might cause such mixtures to be produced simultaneously with arcing or sparking of electrical equipment, or in which electrically conductive dusts may be present	Grain processing Starch processing Starch molding (candy plants) Wood-flour processing	Electrical—Type EX[e] Diesel—Type DY Gasoline, diesel Types D & DS, and LP gas not recommended for this service

TABLE 4-5 *(Continued)*
Recommended Trucks for Various Occupancies
(Source: Factory Mutual System)

Location	Typical Occupancies	Types of Trucks[a] Approved and Listed[b]
Class II, Division 2[c] Locations in which explosive mixtures of combustible dusts are not normally present or likely to be thrown into suspension through the normal operation of equipment, but where deposits of such dust may interfere with the dissipation of heat from electrical equipment or where such deposits may be ignited by arcs or sparks from electrical equipment	Storage and handling of grain, starch or wood flour in bags or other closed containers Grinding of plastic molding compounds in tight systems Feed mills with tightly enclosed equipment	Electrical—Types EE, EX, ES[f] Gasoline—Type GOOIEST[f] Diesel—Type DY, DS[f] LP-Gas—Type LPS[f]
Class III, Division 1[c] Locations in which easily ignitible fibers or materials producing combustible flyings are handled, manufactured, or used	Opening, blending, or carding of cotton or cotton mixtures Cotton gins Sawing, shaping, or sanding areas in woodworking plants Preliminary processes in cordage plants	Electrical—Types EE, EX Diesel—Type DY Gasoline, diesel Types D & DS, and LP gas not recommended
Class III, Division 2[c] Locations in which easily ignitible fibers are stored or handled (except in process of manufacture)	Storage of textile and cordage fibers Storage of excelsior, Kapok, or Spanish moss	Electrical—Types EE, ES, EX, preferred; EE Gasoline—Type GOOIEST Diesel—Type DS, DY LP-Gas—Type LPS

[a] Type G (gasoline), Type D (diesel), and Type LP (LP-Gas) trucks are considered to have comparable fire hazard.
[b] Type GOOIEST (gasoline), Type DS (diesel), and Type LSP (LP-Gas) trucks are considered to have comparable fire hazard.
[c] Hazardous location as classified in the National Electrical Code
[d] Class I, Division 2, Group D only; may be used in Groups A, B, and C areas subject to local authority.
[e] Acceptable for Group G, and for Groups E and F, but subject to special investigation.
[f] Acceptable, but subject to special investigation·

- Look in the direction of travel, especially when backing up and maneuvering.
- Avoid stunt driving and horseplay.
- Make full stops, and be sure it is safe before proceeding at blind corners or passing through doorways; convex mirrors are useful to both trucks and pedestrians at blind corners.
- Maintain a safe distance of at least three truck lengths from other moving trucks; there should be no passing of other trucks at intersec-

tions, blind spots, or other dangerous locations; keep to the right or in identified lanes if possible.

- Avoid quick starts, jerky stops, and quick turns; use caution when on ramps, grades, or inclines.
- Do not use fork trucks as "man-lifts," but use ladders, scaffolding, or equipment designed for lifting people.
- Do not use trucks for any purposes for which they were not intended, such as bumping skids, closing truck or freight car doors, moving heavy objects using jury-rigged equipment, or pushing other trucks; disabled trucks should be moved by towing with a tow bar and safety chain or transported on a lowboy trailer.
- Approach an elevator at right angle to the gate, and stop at least 5 feet from the gate until permitted to proceed; on elevators, operators should shift into park or neutral, set brakes, shut off power, and dismount.
- Drive carefully over bridge plates, which must be properly secured; cross railroad tracks on a diagonal when possible; park at least 8 feet from the nearest rail.
- Before driving onto a vehicle, make sure that the vehicle's brakes are set and its wheels chocked.
- Keep feet and legs inside the operator's cab or guards; do not put any part of the body through the uprights of the truck; keep hands where they cannot be pinched between steering or control levers and stationary objects in tight spaces.
- Before leaving a truck unattended, shift into park or neutral, set brakes, lower load engaging device, shut off power, remove key or pull connector plug, and block wheels if on an incline; do not park in an aisle or doorway; do not obstruct material or equipment that must be accessible.
- Look out for pedestrians, sound horn (in moderation) when approaching people, and proceed with caution. Pedestrians must give way without undue delay.
- Do not permit riders on any part of the truck, including the cab, fork, coupling, or trailer.
- Check loads for stability, including those on truck or trailer, pallet, or skid; objects should be interlaced, tied down, blocked, limited in height, or otherwise kept from being loosened or dislodged.
- Do not idle combustion engines for long periods in spaces with limited ventilation. Carbon monoxide levels must be checked periodically.
- Be careful in maneuvering lift trucks which steer by rear wheels, watching for rear swing.
- Specific driver training should cover procedures for safe turning, U-turns, smooth starts and stops, grades, required clearances in all directions, loading and unloading, fork spread, load handling capacity, stability and balance, and so forth.

4.3.3.5. Elevators

Freight elevators are loaded by hand, handtruck, or industrial truck. Large units can carry industrial trucks or motor vehicles with their loads. One major safety concern is the increase in size and weight of palletized loads and lift trucks over the years and their use in older elevators with limited capacity. Load capacity limits must be strictly observed to prevent elevator failure, including inadequate braking and traction/drive slipping.

Hoistways, machine rooms, pits, and overhead areas must be kept clean, and all equipment must be regularly maintained and inspected. In order to prevent trips and falls, handrails, gates, alarms, and guards are necessary around landings, elevator doors, and hoistway doors and gates. Emergency elevator evacuation procedures must be developed and communicated to personnel.

4.3.3.6. Manual Handling

Handling of containers should follow the principles described in Section 4.3.3.1. Other safety tips are:

- If a package is too heavy to be lifted by one person, get help.
- Before lifting, consider the travel route and holding time, with the possibility of tiring.
- To place a package on a bench or table, first lower it on edge as far back and well supported as possible; release gradually; push back from the front to prevent pinched fingers.
- Secure objects to prevent falling, tipping, or rolling off supports.
- To raise a package above shoulder height, first lift waist high; rest the edge on a ledge or support, shift hands, bend knees, and boost the object by straightening the knees.
- To change direction, lift object to carrying position and turn entire body including feet. Do not twist body.

Boxes and cartons should be lifted by grasping alternate top and bottom corners, and drawing a corner between the legs. Sacks should be grasped at opposite corners. After rising to an erect position, the worker can let the sack rest against his hip and stomach, then swing the sack to one shoulder. The worker should then stoop slightly, and put his hand on his hip so that the sack's weight is distributed to the shoulder, arm, and back. When lowering, the sack should be swung around gently, rested against hip and stomach, and lowered by bending the legs while keeping the back straight.

Larger or heavier containers, such as barrels or drums, require special training because of the variable weight and contents. Some different situations are as follows:

- Upend full drum by two persons without lifting aid: Stand on opposite sides of the drum, face each other, grasp rolled edges at both ends, lift

one end and press down on the other, upend and bring to balance on bottom edge, release grip on bottom edge, and straighten up.
- Overturn full drum by two persons without lifting aid: Provide enough maneuvering room to avoid pinching of hands, stand facing the drum, each grip the top rolled edge with both hands, push with palms on side of drum, balance drum on its bottom edge, step forward a short distance, each move one hand from top edge to bottom edge, ease drum down until it rests on its side.

The use of semiautomated or specially designed operator-assist devices such as hydraulic drum lifters or manipulators allows one person operation and can avoid potentially unsafe manual operations. However, operators must be trained to use these devices and be aware of capacity, speed, and motion limitations.

4.3.4. Shipping Vehicles

OSHA and MTB regulations deal specifically with equipment and work practices at loading docks. These cover dockboards and bridge plates, powered industrial trucks, and the proper packing of hazardous materials. Other items related to the shipping vehicle or trailer are covered below. Proper handling of goods and vehicles is especially important when hazardous materials are shipped [7].

4.3.4.1. Movement of Trailers
The area in front of the loading dock should be large enough for tractor-trailer maneuvering. Traffic should generally move up to and away from the dock in a counterclockwise direction, because it takes less room to maneuver a truck to the left than to the right. It is also easier to back a trailer toward a dock from a counterclockwise direction. If clockwise backing must be performed, an extra 100 feet of space beyond the loading area is recommended. Ideally, a Y-angle or recessed approach should be used to accommodate the maximum truck size and reduce the angle of access.

To allow for proper docking and least risk to personnel, docks should have high intensity lighting, safety surfaces, edge markings, floor drainage, and a canopy over the edge over the dock. Canopies are needed to divert wind, rain, snow, and ice, and they should be sloped to direct runoff away from the trucks. If a depressed driveway is used to create artificial dock height, snow and ice must be cleared to avoid accidents.

Vehicles should be equipped with back-up alarms, to warn personnel in the dock area of truck movement. They can also have backing safety devices such as a contact-sensing rear bumper, which automatically applies the brakes if contact is made with a solid object.

4.3.4.2. Immobilization of Trailers

Regulations require that:

- Truck brakes must be set and wheel chocks used under rear wheels when a trailer is boarded by powered industrial trucks. Wheel chocks and stops prevent the truck from moving while loading or unloading. The chocks or stops should be painted a light or reflective color and should be chained or anchored to the face of the dock.
- Fixed jacks may be used to prevent upending when loading or unloading an uncoupled trailer. A stabilizing jack placed under the fifth wheel plate of a trailer prevents tipping or movement due to uneven ground, defective dolly wheels, or load shifts during operations inside the trailer. Dolly pads, which are simple wooden blocks with handles, are also used to prevent trailers from sinking unevenly into soft ground and tipping. A screw or hydraulic jack can also be used to support the trailer nose to avoid collapse of the landing gear assembly because of dolly metal rust or fatigue, defective struts, or other cause.

Once a trailer is immobilized, dockboards are used to span the distance between the truck and the dock. These dockboards or dock plates, bridge plates, gangplanks, and bridge ramps must be strong enough to carry both static and dynamic loads. They should be rated for at least four times the heaviest expected load. Some dockboards also have built-in drop-locks which secure them to the dock and prevent dislodgement. Also used are hydraulic dock levelers, either truck- or operator-actuated, which can vertically move either the truck or the dock until the interface is level.

All of this equipment must be handled with care. Handling of heavy parts by fork lift truck near the dock edge is dangerous to adjacent personnel.

Dock shelters and seals can minimize gaps between the truck and the dock and so maintain interior temperatures and protect personnel and products from the weather. Air doors or curtains, which blow high-velocity, filtered air downward from above doors and passageways, can control the dock environment and help to prevent accidents caused by restricted passageways.

Over time, trucks can crack or break the lip of the dock, increasing the probability of personnel injury or equipment damage. Therefore, dock lips should be protected against shock by rubber, plastic, or steel bumpers.

Finally, a truck should not be moved until all workers are off the truck or properly seated and protected from injury if the load should shift during transit.

4.3.4.3. Trailer Load Packing

Materials should be packed into trailers in such a way as to avoid the shifting of loads during travel. They can be loaded with little or no space between containers or between containers and walls. Containers can be secured with ropes, straps, chains, or cargo nets to prevent movement. Containers having valves or fittings must be loaded properly to minimize the likelihood of damage in transit.

Containers holding hazardous materials must be fastened. If a container is not permanently attached to the motor vehicle, it must be secured from movement within the vehicle during transportation. Further, during loading or unloading, the vehicle handbrake must be set and precautions taken to prevent movement of the vehicle.

When hazardous substances are shipped together with nonhazardous materials, they must be loaded in a way that provides ready access to the hazardous materials for shifting or removal.

4.3.4.4. Inspection of Shipping Vehicles

Before shipment of hazardous materials, the following items must be checked:

- Chemistry and properties of each material have been identified, the material has been classified according to DOT regulations, and the proper DOT shipping name has been selected.
- Proper packaging compatible with the product has been selected, is authorized in the regulations for transportation of that product, and is marked and labeled according to regulations.
- Shipping documents have been prepared, with certification of propriety of the shipment under regulations.
- Documents include an MSDS for each material.
- Proper placards have been selected and installed on containers and on the shipping vehicle.
- Proper closing and protection devices have been provided on containers.
- Containers have been blocked and braced.
- There is no unlawful combination of hazardous materials in a common or single motor vehicle unit.
- Any damaged, broken, or leaking packages have been removed for proper disposal.

If regulated goods are purchased for resale, the shipper still must assure and certify that the shipped goods meet all DOT regulations. Note also that one may not over-label a shipment; one must fully label hazardous material but may not label unregulated commodities as hazardous material.

In addition to the normal safe material handling procedures, inspect the floor for cracks, breaks, or other weakness before driving a powered vehicle onto the floor of a truck or trailer.

4.3.4.5. Hazardous Cargo Routing

Unless there is no practical alternative, a motor vehicle containing hazardous material should not be operated over routes near heavily populated areas or in tunnels, narrow streets, or alleys. If Class A or B explosives are being transported, specific route requirements are imposed by DOT regulations.

A motor vehicle with hazardous materials may not be parked within 300 feet of an open fire nor driven near such a fire unless safe movement can be assured.

All shipments of hazardous material must be completed without unnecessary delay. Shipments that are refused by the consignee or that cannot be delivered within 48 hours after arrival at their destination must be promptly returned to the shipper or disposed of in some acceptable manner.

4.4. PLANT MODIFICATIONS

Plant modifications are not fundamentally different from maintenance and inspection tasks. Since they can be much more extensive, however, they require a somewhat different approach.

4.4.1. Change Control Program

"Change" as used here refers to a new or fundamentally different system or procedure. Replacement in kind is excluded. All true modifications must be well documented. No modifications should be permitted without prior authorization and follow-up documentation. Extensive or higher levels of change should require higher levels of authorization. A carefully planned and executed Change Control Program must be instituted to prevent the hazards that often arise from quick responses.

The change control program should comprise the following steps:

- Request a Change Order.
- Obtain approval from authorized person(s).
- Issue a Work Order.
- Prepare necessary documents (specifications, drawings, etc.).
- Prepare inspection reports for work performed to verify that correct materials were used and the modification was properly implemented.
- Upon completion of the change, communicate new procedures to the operating staff, update all drawings and equipment data, and forward the revised documentation for secure storage and future reference.
- Audit the document control facilities periodically to determine that information is up-to-date.

4.4.2. Change/Work Authorization

A plant modification usuallly requires management approval because of the associated capital expenditure. The review process should also consider safety aspects and should involve engineers or production personnel. This review should identify and document the following:

- detailed scope of work
- hazards likely to be encountered
- changes in nature or level of hazard due to the modification
- resources required (equipment, tools, and personnel)
- safety measures
  —procedures
  —personal protective equipment
  —supervision and monitoring
- personal responsibilities
- schedule
- requirements for isolation of work area
- effects on modified unit
- effects on other operating units
  —during construction
  —after startup
- new demands on safety systems
- compatibility of new components with existing system

The program should be approved by the authorized manager. The scope of the work should be limited to that described in the authorized work plan; there are to be no "blank checks." Specific activities may require individual approval by formal permits. These are discussed in Section 7.1.4.

4.4.3. Training

Plant personnel should immediately be made aware that a plant modification has been made. Its effects on operating and maintenance practices must be clear. Management must establish:

- need for formal training
- extent of training required
- affected personnel
  —supervisors
  —operators
  —maintenance staff
- timing

Personnel should be trained before actually working with the modified equipment or procedure. Training is discussed in greater detail in Chapter 6.

4.5. HAZARDOUS WORK

This section describes the precautions necessary with certain types of hazardous work. It covers entry into confined spaces, lockout and tagging of equipment for maintenance, opening of closed vessels and pipelines, work with highly hazardous materials, and welding. Under certain conditions, such work will require a permit. The subject of permits is discussed in Section 7.1.4.

4.5.1. Confined Space Entry

Examples of confined spaces whose entry may involve extra hazards are pits, dikes, excavations, control panels, sewers, crowded areas, and vessels. The hazards frequently encountered in confined spaces are

- oxygen deficiency
- presence of toxic, flammable, or asphyxiating gases
- entry of hazardous material during the work
- presence of rotating equipment next to personnel
- need to work in cramped or unnatural postures
- tripping hazards or poor footing

Asphyxiation, due either to an oxygen deficiency or to a toxic atmosphere, is the leading cause of death in confined spaces [8]. NIOSH [9] reports the striking fact that most of these fatalities occur among would-be rescuers. Proper training and the use of adequate equipment are essential if such incidents are to be avoided. All employees should be trained in the hazards of improper entry and should be aware that the natural impulse to rush to someone's rescue without some preparation may only aggravate the situation and may put the rescuer also at risk. The chemical hazards are overcome by removing dangerous substances and isolating the confined space to prevent entry of hazardous material.

OSHA is heavily involved in the regulation of entry into confined spaces and actually mandates permit procedures. The requirements are covered in detail in Section 7.1.4. An OSHA "confined space" is defined in 29CFR1910.146 as having three characteristics:

- large enough and so configured that an employee can bodily enter and perform assigned work
- limited or restricted means for entry or exit
- not designed for continuous occupancy

An entry permit is required whenever a confined space has any of the following characteristics:

- contains or has the potential to contain a hazardous atmosphere
- contains a material that has the potential for engulfing an entrant
- has an internal configuration such that an entrant could be asphyxiated or trapped by inwardly converging walls or by a sloping floor which tapers to a smaller cross-section
- contains any other recognized serious safety or health hazard.

The acceptable level of a hazardous vapor will be a function of its LEL or MAC. Atmospheric testing should be done only with tested and approved equipment. The test apparatus should be calibrated frequently or immediately before use. When the work itself generates toxic or flammable vapor (e.g., welding), or when there is the possibility that such vapors will enter the space, continual monitoring is necessary.

Oxygen concentrations should be at least 19.5%, unless air-supplied respiratory equipment is used. Where possible, positive ventilation should be used and continued as long as anyone remains in the confined space. Air should be supplied in such a way and at such a location that it must pass the worker(s) before exhausting.

If work ceases for any reason, the atmospheric testing should be repeated before reentry.

The equipment provided for use by backup personnel should be covered by the maintenance manual and work permits. Such equipment includes the protective equipment necessary for entry and the gear required for rescue of personnel from the confined space. The latter includes chain falls, lifelines, and harnesses. These devices must be securely fastened outside the space. The best choice will depend on the shape of the access port(s). In the case of a closed vessel provided with one or more manways, for example, a wrist harness may be preferable to a full-body harness. The latter may make it very difficult to remove a person who is badly injured or unconscious.

Communication is important. If the work location is inherently noisy, there must be special provision for communication between workers and standby personnel. Communication can be electronic or, for most standard situations, nonverbal (e.g., hand signals, horn, lights). There should also be communication between standby personnel and others not involved in the maintenance work, in case more help is suddenly needed. This communication can be through an electronic link or can be as simple as an agreed-upon trouble alarm.

4.5.2. Equipment Lockout

Whenever the inadvertent or unauthorized starting of a piece of driven equipment will create a hazard for inspection or maintenance personnel, it is important to disable that equipment while work is being performed. The hazard might be strictly mechanical, as in contact with rotating machinery, or partly chemical, as in a release caused by the pressure developed by a pump or compressor.

If its driver is a motor, the equipment is taken off line by opening a switch or breaker and locking it to prevent closure. The switch should be individually locked or tagged by each person or department involved with the work. At a minimum, this would include the direct operator of the equipment and the inspecting or maintenance group. The first lock applied, and the last removed, should be that of the organization responsible for the equipment. When the switch or breaker is first locked out, the production supervisor or operator should attempt to start the piece of equipment in question. This will verify that the correct switch has been disabled.

The maintenance or inspection group should check with production before beginning work and should add their tags to the switch as appropriate. Each tag should be signed and dated. None should be removed except by the person who affixed it (or a delegate). Tags should be removed only when the responsible party is convinced that all work has been performed properly. Figure 4-7 shows examples of tags proposed by OSHA.

When the drive is a steam turbine, similar precautions are necessary. The drive should be disabled by locking (if possible) and tagging steam valves, breaking or blocking the steam supply line (see Section 4.5.3, below), or disconnecting the transmission between the turbine and the driven equipment. Each organization involved in the work being performed should review arrangements, as above, and apply tags if appropriate.

4.5.3. Line Breaking and System Opening

The work authorization procedures discussed in Section 4.4 must be satisfied before equipment or piping is opened for inspection or maintenance. The contents of the system and the appropriate safety measures must also be identified. Before work begins, maintenance personnel should verify the following:

- Work has been authorized and necessary permits have been granted.
- All requirements for apparatus, tools, protective equipment, replacement parts, and standby personnel have been fulfilled.

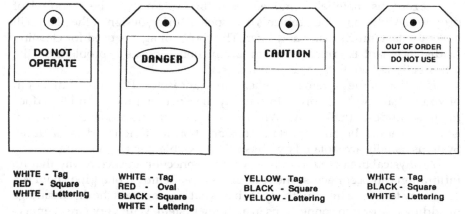

FIGURE 4-7. Accident-Prevention Tags. (OSHA recommendation; Reference 29 CFR 1926.200, Subpart G.)

- Material that may be contained in the line or equipment has been identified. (Never assume complete removal by flushing, venting, or draining; consider the possibility of concentration of slightly soluble or nonvolatile impurities.)
- Accumulated pressure has been bled from hydraulic drives.
- Hazardous properties of the materials known or suspected to be present have been identified.
- Problems resulting from residual pressure in the system have been considered and guarded against.
- Countermeasures against possible release are in place:
  —collection
  —destruction
  —containment
  —local ventilation
  —evacuation of area
  —shutdown of nearby systems

The first priority is the removal of hazardous material from the system. This can be done by purging, flushing, or mechanical or chemical cleaning. The last of these methods is treated from a process standpoint in Chapter 5. Flushing usually involves the displacement or dissolution of material by water or another solvent. This is also an example or extension of the chemical cleaning described in Chapter 5. Purging is accomplished by displacing process vapors with less hazardous material. This can be done continuously or by repeated addition and removal of the chosen safe gas. The latter approach dilutes the hazardous material in stages until its concentration is deemed acceptable. In either case, the purge gas or contents of the system should be sampled and analyzed to verify results before breaking or opening. Inertion is covered in more detail in Chapter 2.

Hazardous materials removed from the system must be disposed of properly. Addressing the problems of disposal during design of the plant will make the task much easier and safer. The choice of agents used for cleaning or displacement is critical; it is important not to create a problem by the introduction of another hazardous material.

Isolation of the system is another important factor. The minimum length or volume should be involved to simplify the isolation process and to reduce the magnitude of the hazard. When very hazardous materials are handled, isolation should be an important consideration in design and installation. Systems usually are isolated by physical breaks, blanking, or valving.

A physical break involves removing a connection in such a way that no infiltration can occur across the resulting space. Removal of a pipe spool is a good way to ensure this, but the spool has what is sometimes the disadvantage of adding one or two connections in a piping system. With very large connections (e.g., hoppers, conveyor feed ports, vent or relief systems, other attached equipment), a physical break may not be practical. In such a case, careful inspection and monitoring, combined with the use of blinds, equipment lockout, and isolation or placarding of an area, can still provide safe working conditions.

Blanking is essentially an elaboration on the preceding method. Blanks are usually acceptable for low-pressure air and water service or for negative pressures. In more demanding service, the blanks must have the strength required to withstand full normal line pressure and must be of a material compatible with the fluid carried by the line.

Valving is frequently used, but should not rely on a single valve for security. The integrity of such a valve should always be questioned. Double valving can improve security against leakage and is often supplemented by providing an opening to atmosphere between two block valves (double-block-and-bleed or valve-vent-valve arrangement). Valves should be locked and tagged in accordance with Section 4.5.2.

4.5.4. Hazardous Materials

Areas containing hazardous materials require extra measures to protect workers involved in installation, renovation, or maintenance work. Any work in such areas should be preceded by a work plan identifying

- the work objectives
- the methods used to achieve those objectives (e.g., welding, sampling, nondestructive testing)
- the documentation that should be readily available before starting the work:
  —operating procedures
  —system drawings
  —equipment layout drawings
  —piping, electrical, instrumentation, ventilation duct, and structural steel drawings
  —MSDS for hazardous materials in the area
- the hazardous areas affected by the work, a plot plan showing the boundaries of those areas, and the procedures for restricting access to the areas
- construction materials and spare parts and when they will be needed.
- personnel protection requirements:
  —personal protective equipment: clothing, eye protection, gas masks, etc.
  —portable ventilation and air circulation equipment.
  —fire watches and fire protection equipment.

- qualifications and training required for the workers performing the tasks
- names of key personnel and their alternates
- schedule of activities which includes all precautions to isolate or eliminate hazardous materials from the work area
- means of notifying operating personnel in the affected areas of the work's starting time and duration
- list of standard procedures for the work to be performed, together with a checklist that can be marked off as each task is completed
- cleanup and decontamination procedures complete with a checklist

The work should not begin until all necessary protective equipment is readily available to the personnel performing the work.

4.5.5. Welding

Welding is the most common ignition source in many plants. The safety precautions to be taken when welding are given below. General precautions are first, followed by those which apply especially to gas welding and to arc welding. A discussion of the hazards of welding and cutting in process areas and a list of applicable codes and standards are also included.

A. General Welding Precautions
- Make sure the work area is well ventilated. As a minimum, a portable exhaust hood should be placed over the welding area.
- Be sure that personnel are adequately protected with fire-resistant gloves, overalls, jackets, aprons, goggles, etc.
- Place curtains around the work area to protect nearby personnel from sparks, hot debris, and the glare of a torch or arc.
- Use ear protection in areas where noise is a problem.
- Be sure that only adequately trained and qualified personnel who are familiar with the equipment being used perform the welding.
- Stop all work associated with the handling or transfer of flammable liquids in the area.

B. Gas Welding
- Never move a cylinder by dragging, sliding, or rolling it. Cylinders should be vertically mounted and chained on a truck to prevent damage or abuse.
- Never permit grease or oil to come in contact with cylinder valves.
- Never place a cylinder near a heat source such as a furnace or torch sparks.
- Never store or leave cylinders in a confined area.
- Never place a cylinder in a horizontal position.
- Never remove the valve protector cap before using.

- Never attempt to repair leaking or malfunctioning cylinder valves. Return to supplier.
- Never use oxygen as a substitute for compressed air.
- When opening cylinder valves, always stand to the side and away from the regulator.

C. Arc Welding

- Use only welding equipment that has been constructed and installed in accordance with the National Electrical Code (NEC).
- Locate the power disconnect switch near the welder for quick and convenient shut-off.
- Be sure the welder is shut off before attempting any repairs to the machine.
- Keep the welder clean and cool.
- Do not leave the electrode grounded to the work for prolonged periods.
- Be sure the welder is properly grounded to the work to prevent shocks from stray currents.
- Never attempt to operate the welder beyond its rated capacity.
- Never change the polarity of the welder while it is in operation.
- Be sure that the welding cables are in good condition. To prevent overload and overheating, do not use longer cables than recommended.
- Avoid working in damp areas. If this is unavoidable, keep hands and clothing dry at all times.
- Do not strike an arc until welding curtains are placed around the area and personnel in the area are wearing adequate eye protection.

D. Welding and Cutting Piping, Containers, and Vessels

- No welding or cutting should be performed unless all combustible and toxic material has been thoroughly removed from the piping, container, or vessel.
- The method of cleaning and the materials used for cleaning should be established in the work plan. Some of the methods that are usually considered are:
  —Cleaning with a water-soluble solvent or detergent.
  —Hot chemical cleaning followed by a repeated water or steam flush. Hot or compressed air is sometimes used in the final step in the cleaning sequence.
  —Steam cleaning alone or after flushing the system with a caustic cleaning solution.
- A useful precaution when welding or cutting a container that once held volatile materials is to fill the container with water to within a few inches of where the welding is to be performed.
- Maintain a fire watch during welding and cutting, observing whether sparks or flying slag may fall near flammable material or unprotected

personnel. Have fire extinguishing equipment nearby. Have fire resistant guards and screens around the work area.

- When working in an area with significant dust or vapor loading, establish a work plan that reduces the chance of explosions from welding arcs and flames, such as using portable vent hoods over the work area.
- If welding or cutting over a wooden floor, sweep floor clean and wet down. Perform welding over a bucket or pan containing water or sand to collect dropping slag.
- When possible, work in an open area to reduce the hazards from flying sparks and slag.

E. Codes and Standards

Following are some of the codes and standards that serve as guides to safe welding practices:

ANSI/NFPA	51B Standards for Fire Prevention in Use of Cutting and Welding Processes
ANSI/UL 591-1980	Safety Standards for Transformer-Type Arc-Welding Machines
API Spec. 12D	Specification for Field-Welded Tanks for Storage of Production Liquids (with supplements)
API Spec. 12F	Specification for Shop-Welded Tanks for Storage of Production Liquids (with supplements)
API Std 1104	Standards for Welding Pipelines and Related Facilities
API Std. 620	Recommended Rules for Design and Construction of Large, Welded, Low-Pressure Storage Tanks
API Std. 650	Welded Steel Tanks for Oil Storage Tanks
AWS D10.9	Specifications for Qualification of Welding Procedures and Welders for Piping and Tubing
AWS F4.1	Recommended Safe Practices for Welding and Cutting Containers and Piping That Have Held Hazardous Substances
AWS/ASC Z49.1	Safety in Welding and Cutting

4.6. OUTSIDE CONTRACTORS

Outside contractors are used for several reasons:

- to provide expertise which may be lacking in the operating company
- to handle specific projects for which the operating company does not have sufficient manpower for timely execution
- to provide routine services such as engineering and maintenance

In any of these cases, there will be a contract between the two parties and perhaps the use of certain subcontractors. This section discusses some of the safety aspects of such contracts. It then turns to methods which can be used to assure that the contractor has the tools and the ability to perform safely in the plant.

4.6.1. Contract Language

Contract language is important in setting the responsibility of the contractor to work in a safe manner. Specific language should be included in the contract to define this responsibility. At the same time, the owner of the plant has the obligation under Responsible Care® (Section 7.3) and Title 29 OSHA regulations to provide all necessary information and certain training and to monitor the contractor's performance.

Contract language should require all precautions necessary to the safety of personnel and the enforcement of all laws and plant rules and regulations. All contractors should be made responsible for their performance in these areas. Besides performing safely while executing the work, the contractor should also be obliged to function in a workmanlike manner so that the facility will operate safely after startup. The contractor should warrant that all services show the degree of care and skill expected in the engineering and construction industry.

The above does not set specific criteria for performance. Such clauses are sometimes applied to large projects and always entail the added problem of agreeing to methods and techniques for measuring results. It likewise does not set liabilities. These can be set in the contract, but the contractor will quite properly seek to limit them.

4.6.2. Pre- and Postcontract Meetings

Meetings held before the signing of a contract should address safety issues as conscientiously as they address financial issues. The contractor's safety record and program should be among the criteria for selection. These meetings should specifically address the ways in which the contractor's and the owner's safety programs will be merged for best results. The owner is obliged to see that contractor employees are given training equivalent to that received by owner employees. This can be done simply by exposing both groups to the same training and is easily accomplished when an ongoing service contract is in place and contractor employees are continuously present on the plant.

After completion of a contract, other meetings usually are held to review performance and to capture for future application the lessons learned during the contract. Safety should be one of the topics of these meetings. The contractor and owner should jointly identify both successes and failures and list their recommendations for improvement.

4.6.3. Contractor Safety Programs

Section 4.6.1 suggested that the contract should contain language requiring the contractor to perform his work safely. This sometimes can be achieved by including the contractor in the plantwide safety program. An alternative would be to have him operate under his own safety program, subject to its review and approval by the plant's safety department. The contractor's safety program should include at least the following:

- information pertinent to the recognition and avoidance of unsafe conditions and the regulations applicable to the work environment
- instructions regarding the safe handling and use of poisons with which workers may come in contact
- information on hazards, protective measures, and personal hygiene
- directions for safe handling and use of flammable and toxic materials
- procedures for work in confined or enclosed spaces
- methods for documenting and reporting safety violations

A typical contractor safety manual should comprise the following:

- Accident prevention organization
- Personal protection
- Housekeeping
- Sanitation
- First aid
- Fire prevention and protection
- Electrical installations
- Pipelines
- Handling and storage of materials
- Flammable gases and liquids
- Explosives
- Ladders
- Scaffolding
- Hoists, cranes, and derricks
- Heavy equipment
- Motor vehicles
- Barricades
- Excavation and shoring
- Demolition
- Pile driving
- Welding and cutting
- Steel erection
- Concrete construction
- Masonry
- Hand tools
- Power tools

- Power-actuated tools
- Boilers
- Pressure vessels
- Safety procedures forms
  —Emergency telephone numbers
  —Safety inspection checklist
  —Record of accident
  —Record of first aid treatment

4.6.4. Monitoring

OSHA has the right of entry to any site to inspect or investigate for compliance with federal regulations. This does not reduce the responsibility of the plant owner, who should, at a minimum, monitor contractor incident reports for adverse trends. The owner should also have a program of surveillance by which he can monitor the contractor's day-to-day adherence to plant safety rules.

4.6.5. Training

When a contract is let, it is necessary to determine the extent of training required for contractor personnel. The plant owner must establish:

- need for training
- extent and type of training
- administration of training
  —contractor
  —plant personnel
- audience
  —supervisors
  —technicians
  —laborers
  —craftsmen
- timing
- location

Training is discussed in greater detail in Chapter 6.

4.6.6. HAZCOM and Hazard Analysis

OSHA (29CFR1910.1200) requires that hazards from all materials produced or imported to the plant be evaluated and that this hazard information be disseminated to all affected personnel. This should include not only plant personnel but also employees of outside contractors. By the same token, the outside contractor must communicate in advance the hazards from materials imported to the site during the execution of the contractor's services. The use of MSDS and labeling for hazard communication is described in Chapter 2.

Before any plant modification or installation of a new process system, the plant must perform a study to identify the hazards that could result from this installation. Section 4.4.2 discusses management of change. The outside contractor can be invited to contribute to this study and to the hazard analysis. Section 7.1.1 mentions some of the various approaches to hazard analysis.

4.6.7. Enforcement

Enforcement of plant safety requirements occurs at several levels. OSHA can levy fines and demand remedial action to mitigate any conditions adverse to safety. These actions are taken against the plant owner who has the ultimate responsibility for all incidents at his facility. Considering the consequences of unsafe operations, the plant operator may choose to tie payments to the safety performance of the contractor.

4.7. WORKER PROTECTION

4.7.1. Personal Protective Equipment

This section covers the types of personal protective equipment (PPE) appropriate to various situations. PPE is not to be considered a primary defense against hazards. Direct hazard control or administrative and engineering controls must come first. PPE then is used to back up or supplement these primary measures.

PPE protects the respiratory system, skin, eyes, face, hands, feet, head, body, or hearing from chemical, physical, and biological hazards. No single combination of protective clothing and equipment can protect a person from all hazards. To be effective, the selection and use of PPE must be appropriate and accompanied by adequate training. Each plant should have a PPE program which includes

- hazard identification
- selection, use, and maintenance of PPE
- decontamination procedures
- training
- medical monitoring
- environmental surveillance
- identification of hazards associated with the PPE itself (e.g., heat stress, restricted vision and mobility)

Major categories of PPE are considered below and are described in a series of tables and illustrations patterned after those issued by DHHS [10]. OSHA regulations will be found in the following subsections of 29CFR1910:

Information relating to PPE used in construction work is in Section 1926.28 of the same Title.

4.7.1.1. Respiratory Equipment

Respirators may supply air directly to a user (atmosphere-supplying type) or remove contaminants from the ambient air (air-purifying type). Figure 4-8 illustrates these configurations, and Table 4-6 lists their advantages and disadvantages. Table 4-7 compares the different types of self-contained breathing apparatus (SCBA). Each type of respirator is in some way limited in the protection it affords. SCBA has a finite supply of air. Air purifiers are of no use in oxygen-deficient atmospheres and also have limited capacities. Supplied-air respirators (SAR) may require supplemental SCBA and limit mobility. Employees must be aware of these limitations and able to choose the right equipment for a given hazard.

4.7.1.2. Protective Clothing and Accessories

Protective clothing includes

- fully encapsulating suits
- nonencapsulating suits
- fire protective suits
- proximity or approach clothing
- blast and fragmentation suits
- cooling suits
- radiation protective suits
- aprons, leggings, sleeve protectors
- gloves

Figure 4-9 shows two forms of protective clothing, and Table 4-8 lists protective clothing and accessories for various parts of the body.

4.7.1.3. Complete Outfits

Different hazards require different protective equipment ensembles. OSHA has defined four levels of protection in 29CFR1910.120, App. B. Table 4-9 lists typical combinations for each level and gives criteria for selection and limitations on use. There are preferred donning and doffing procedures for these outfits, as well as storage and monitoring methods.

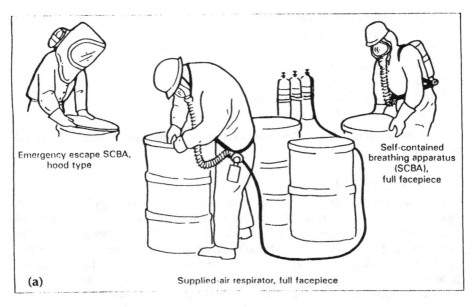

Emergency escape SCBA, hood type

Self-contained breathing apparatus (SCBA), full facepiece

(a)   Supplied-air respirator, full facepiece

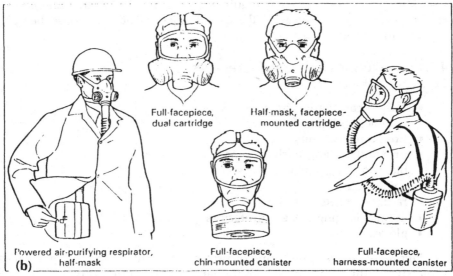

Full-facepiece, dual cartridge

Half-mask, facepiece-mounted cartridge.

Powered air-purifying respirator, half-mask

(b)

Full-facepiece, chin-mounted canister

Full-facepiece, harness-mounted canister

FIGURE 4-8. Types of Respirators. (a) Atmosphere-supplying regulators. (b) Air-purifying respirators.

TABLE 4-6
Relative Advantages and Disadvantages of Respiratory Protective Equipment

Type of Respirator	Advantages	Disadvantages
ATMOSPHERE-SUPPLYING		
Self-Contained Breathing Apparatus (SCBA)	' Provides the highest available level of protection against airborne contaminants and oxygen deficiency ' Provides the highestavailable level of protection under strenuous work conditions	' Bulky, heavy (up to 35 lb) ' Finite air supply limits work duration ' May impair movement in confined spaces
Positive-Pressure Supplied-Air Respirator (SAR) (also called air-line respirator)	' Enables longer work periods than an SCBA ' Less bulky and heavy than an SCBA. SAR equipment weighs less than 5 pounds (or around 15 lb if escape SBCA protection is included)	' Not approved for atmospheres immediately dangerous to life or health (IDLH) or in oxygen-deficient atmospheres unless equipped with an emergenct egress unit such as an escape-only SCBA that can provide immediate emergency respiratory protection in case of air-line failure. ' Impairs mobility ' MSHA/NIOSH certification limits hose length to 300 ft (90 m) ' As the length of the hose increasesm the minimum approved air flow may not be delivered at the facepiece ' Air line is subject to damage, chemical contamination, and degradation. Decontamination of hoses may be difficult. ' Worker must retrace steps to leave work area ' Requires supervision/monitoring of the air supply line
AIR-PURIFYING		
Air-purifying Respirator (including powered air-purifying respirators [PAPRs])	' Enhanced mobility ' Lighter in weight than an SCBA. Generally eighs 2 lb (1 kg) or less (except for PAPRs)	' Cannot be used in IDLH or oxygen-deficient atmospheres (less than 19.5% O_2 at sea level) ' Limited duration of protection. May be hard to gauge safe operating time in field conditions. ' Only protects against specific chemicals and up to specific concentrations. ' Use requires monitoring of contaminant and O_2 levels ' Can only be used (1) against gas and vapor contaminants with adequate warning properties, or (2) for specific gases or vapors provided that the service is known and a safety factor is applied ofr if the unit has an ESLI (end-of-service-life indicator)

TABLE 4-7
Types of Self-Contained Breatining Apparatur (SCBA)

Description	Advantages	Disadvantages	Comments
ENTRY-AND-ESCAPE SCBA Open-Circuit SCBA			
Supplies clean air to the wearer drom a cylinder. Wearer exhales air directly to the atmosphere.	Operated in a positive-pressure mode, open-circuit SCBAs provide the highest respiratory protection currently available. A warning alarm signals when only 20–25% of the air supply remains	Shorter operating time (30–60 min) and heavier weight (up to 35 lb [15.9 kg]) than a closed-circuit SCBA	The 30–60 min operating time may vary depending on the size ot the air tank and the work rate of the individual
Closed-circuit SCBA (Rebreather)			
These devices recycle exhaled gases (CO_2, O_2, and N_2) by removing CO_2 with an alkaline scrubber and replenishing the consumed O_2 from a liquid or gaseous source	Longer operating time (up to 4 hr) and lighter weight (21–30 lb [9.5–13.6 kg] than open-circuit apparatus A warning alarm signals when only 20–25% of the O_2 supply remains O_2 supply is depleted before the CO_2 sorbent scrubber supply, thereby protecting the wearer from CO_2 breakthrough	At very cold temperatures, scrubber efficiency may be reduced and CO_2 breakthrough may occur Units retain heat normally exchanged din exhalation and generate heat in CO_2 scrubbing, adding to the danger of heat stress. Auxiliary cooling devices may be required. When worn outside an encapsulating suit, the breathing bag may be permeated by chemicals , contaminating the apparatus and the respirable air. Decontamination of the breathing bag may be difficult	Positive-pressure closed-circuit SCBAs offer substantially more protection than negative-pressure units, which are not recommended on hazardous waste sites. While these devices may be certified as closed-circuit SCBAs, NIOSH cannot certify closed-circuit SCBAs as positive-pressure devices due to limita-tions in certification procedures currently defined in 30 CFR Part 11
ESCAPE-ONLY SCBAs			
Supplies clean air to the wearer from either an air cylinder or from an O_2-gener-ating chemical. Approved for escape purposes only	Light weight (10 lb [4.5 kg] or less), low bulk, easy to carry. Available in pressure-demand and continuous-flow models	Cannot be used for entry.	Provides only 5–15 minutes of respiratory protection, depending on the model and the breathing rate of the wearer.

Fully-encapsulating suit

Apron, gloves, hardhat,
faceshield, boot covers

FIGURE 4-9. Examples of Protective Clothing,

TABLE 4-8
Protective Clothing and Accessories

BODY PART PROTECTED	TYPE OF CLOTHING OR ACCESSORY	DESCRIPTION	TYPE OF PROTECTION	USE CONSIDERATIONS
Full Body	Fully-encapsulating suit	One-piece garment. Boots and gloves may be integral, attached and replaceable, or separate.	Protects against splashes, dust, gases, and vapors.	Does not allow body heat to escape. May contribute to heat stress in wearer, particularly if worn in conjunction with a closed-circuit SCBA; a cooling garment may be needed. Impairs worker mobility, vision, and communication.
	Non-encapsulating suit	Jacket, hood, pants, or bib overalls, and one-piece coveralls.	Protects against splashes, dust, and other materials but not against gases and vapors. Does not protect parts of head or neck.	Do not use where gas-tight or pervasive splashing protection is required. May contribute to heat stress in wearer. Tape-seal connections between pant cuffs and boots and between gloves and sleeves.
	Aprons, leggings, and sleeve protectors	Fully sleeved and gloved apron. Separate coverings for arms and legs. Commonly worn over non-encapsulating suit.	Provides additional splash protection of chest, forearms, and legs.	Whenever possible, should be used over a non-encapsulating suit (instead of using a fully-encapsulating suit) to minimize potential for heat stress. Useful for sampling, labeling, and analysis operations. Should be used only when there is a low probability of total body contact with contaminants.

Firefighters' protective clothing	Gloves, helmet, running or bunker coat, running or bunker pants (NFPA No. 1971, 1972, 1973), and boots.	Protects against heat, hot water, and some particles. Does not protect against gases and vapors, or chemical permeation or degradation. NFPA Standard No. 1971 specifies that a garment consist of an outer shell, an inner liner, and a vapor barrier with a minimum water penetration of 25 lbs/in² (1.8 kg/cm²) to prevent the passage of hot water.	Decontamination is difficult. Should not be worn in areas where protection against gases, vapors, chemical splashes, or permeation is required.
Proximity garment (approach suit)	One- or two-piece overgarment with boot covers, gloves, and hood of aluminized nylon or cotton fabric. Normally worn over other protective clothing, such as chemical-protective clothing, firefighters' bunker gear, or flame-retardant coveralls.	Protects against brief exposure to radiant heat. Does not protect against chemical permeation or degradation. Can be custom-manufactured to protect against some chemical contaminants.	Auxiliary cooling and an SCBA should be used if the wearer may be exposed to a toxic atmosphere or needs more than 2 or 3 minutes of protection.
Blast and fragmentation suit	Blast and fragmentation vests and clothing, bomb blankets, and bomb carriers.	Provides some protection against very small detonations. Bomb blankets and baskets can help redirect a blast.	Does not provide hearing protection.

241

BODY PART PROTECTED	TYPE OF CLOTHING OR ACCESSORY	DESCRIPTION	TYPE OF PROTECTION	USE CONSIDERATIONS
Full Body (cont.)	Radiation-contamination protective suit	Various types of protective clothing designed to prevent contamination of the body by radioactive particles.	Protects against alpha and beta particles. *Does NOT protect against gamma radiation.*	Designed to prevent skin contamination. If radiation is detected on site, consult an experienced radiation expert and evacuate personnel until the radiation hazard has been evaluated.
	Flame/fire retardant coveralls	Normally worn as an undergarment.	Provides protection from flash fires.	Adds bulk and may exacerbate heat stress problems and impair mobility.
	Flotation gear	Life jackets or work vests. (Commonly worn underneath chemical protective clothing to prevent flotation gear degradation by chemicals.)	Adds 15.5 to 25 lbs (7 to 11.3 kg) of buoyancy to personnel working in or around water.	Adds bulk and restricts mobility. Must meet USCG standards (46 CFR Part 160).
	Cooling garment	One of three methods: (1) A pump circulates cool dry air throughout the suit or portions of it via an air line. Cooling may be enhanced by use of a vortex cooler, refrigeration coils, or a heat exchanger. (2) A jacket or vest having pockets into which packets of ice are inserted. (3) A pump circulates chilled water from a water/ice reservoir and through circulating tubes, which cover part of the body (generally the upper torso only).	Removes excess heat generated by worker activity, the equipment, or the environment.	(1) Pumps circulating cool air require 10 to 20 ft³ (0.3 to 0.6 m³) of respirable air per minute, so they are often uneconomical for use at a waste site. (2) Jackets or vests pose ice storage and recharge problems. (3) Pumps circulating chilled water pose ice storage problems. The pump and battery add bulk and weight.

Head	Safety helmet (hard hat)	For example, a hard plastic or rubber helmet.	Protects the head from blows.	Helmet shall meet OSHA standard 29 CFR Part 1910.135.
	Helmet liner		Insulates against cold. Does not protect against chemical splashes.	
	Hood	Commonly worn with a helmet.	Protects against chemical splashes, particulates, and rain.	
	Protective hair covering		Protects against chemical contamination of hair. Prevents the entanglement of hair in machinery or equipment. Prevents hair from interfering with vision and with the functioning of respiratory protective devices.	Particularly important for workers with long hair.
Eyes and Face[a]	Face shield	Full-face coverage, eight-inch minimum.	Protects against chemical splashes. Does not protect adequately against projectiles.	Face shields and splash hoods must be suitably supported to prevent them from shifting and exposing portions of the face or obscuring vision. Provides limited eye protection.

[a]All eye and face protection must meet OSHA standard 29 CFR Part 1910.133.

TABLE 4-8
Protective Clothing and Accessories

BODY PART PROTECTED	TYPE OF CLOTHING OR ACCESSORY	DESCRIPTION	TYPE OF PROTECTION	USE CONSIDERATIONS
Eyes and Face (cont.)	Splash hood		Protects against chemical splashes. Does not protect adequately against projectiles.	
	Safety glasses		Protect eyes against large particles and projectiles.	If lasers are used to survey a site, workers should wear special protective lenses.
	Goggles		Depending on their construction, goggles can protect against vaporized chemicals, splashes, large particles, and projectiles (if constructed with impact-resistant lenses).	
	Sweat bands		Prevents sweat-induced eye irritation and vision impairment.	
Ears	Ear plugs and muffs		Protect against physiological damage and psychological disturbance.	Must comply with OSHA regulation 29 CFR Part 1910.95. Can interfere with communication. Use of ear plugs should be carefully reviewed by a health and safety professional because chemical contaminants could be introduced into the ear.

244

Body Area	Equipment	Description	Purpose	Notes
	Headphones	Radio headset with throat microphone.	Provide some hearing protection while enabling communication.	Highly desirable, particularly if emergency conditions arise.
Hands and Arms	Gloves and sleeves	May be integral, attached, or separate from other protective clothing.	Protect hands and arms from chemical contact.	Wear jacket cuffs over glove cuffs to prevent liquid from entering the glove. Tape-seal gloves to sleeves to provide additional protection.
		Overgloves.	Provide supplemental protection to the wearer and protect more expensive undergarments from abrasions, tears, and contamination.	
		Disposable gloves.	Should be used whenever possible to reduce decontamination needs.	
Foot	Safety boots	Boots constructed of chemical-resistant material.	Protect feet from contact with chemicals.	All boots must at least meet the specifications required under OSHA 29 CFR Part 1910.136 and should provide good traction.
		Boots constructed with some steel materials (e.g., toes, shanks, insoles).	Protect feet from compression, crushing, or puncture by falling, moving, or sharp objects.	
		Boots constructed from nonconductive, spark-resistant materials or coatings.	Protect the wearer against electrical hazards and prevent ignition of combustible gases or vapors.	

245

TABLE 4-8
Protective Clothing and Accessories

BODY PART PROTECTED	TYPE OF CLOTHING OR ACCESSORY	DESCRIPTION	TYPE OF PROTECTION	USE CONSIDERATIONS
Foot (cont.)	Disposable shoe or boot covers	Made of a variety of materials. Slip over the shoe or boot.	Protect safety boots from contamination. Protect feet from contact with chemicals.	Covers may be disposed of after use, facilitating decontamination.
General	Knife		Allows a person in a fully-encapsulating suit to cut his or her way out of the suit in the event of an emergency or equipment failure.	Should be carried and used with caution to avoid puncturing the suit.
	Flashlight or lantern		Enhances visibility in buildings, enclosed spaces, and the dark.	Must be intrinsically safe or explosion-proof for use in combustible atmospheres. Sealing the flashlight in a plastic bag facilitates decontamination. Only electrical equipment approved as intrinsically safe, or approved for the class and group of hazard as defined in Article 500 of the National Electrical Code, may be used.

Personal dosimeter		Measures worker exposure to ionizing radiation and to certain chemicals.	To estimate actual body exposure, the dosimeter should be placed inside the fully-encapsulating suit.
Personal locator beacon	Operated by sound, radio, or light.	Enables emergency personnel to locate victim.	
Two-way radio		Enables field workers of communicate with personnel in the Support Zone.	
Safety belts, harnesses, and lifelin-		Enable personnel to work in elevated areas or enter confined areas and prevent falls. Belts may be used to carry tools and equipment.	Must be constructed of spark-free hardware and chemical-resistant materials to provide proper protection. Must meet OSHA standards in 29 CFR Part 1926.104.

247

TABLE 4-9
Sample Protective Ensembles

LEVEL OF PROTECTION	EQUIPMENT	PROTECTION PROVIDED	SHOULD BE USED WHEN:	LIMITING CRITERIA
A	RECOMMENDED: • Pressure-demand, full-facepiece SCBA or pressure-demand supplied-air respirator with escape SCBA. • Fully-encapsulating, chemical-resistant suit. • Inner chemical-resistant gloves. • Chemical-resistant safety boots/shoes. • Two-way radio communications. OPTIONAL: • Cooling unit. • Coveralls. • Long cotton underwear. • Hard hat. • Disposable gloves and boot covers.	The highest available level of respiratory, skin, and eye protection.	• The chemical substance has been identified and requires the highest level of protection for skin, eyes, and the respiratory system based on either: — measured (or potential for) high concentration of atmospheric vapors, gases, or particulates or — site operations and work functions involving a high potential for splash, immersion, or exposure to unexpected vapors, gases, or particulates of materials that are harmful to skin or capable of being absorbed through the intact skin. • Substances with a high degree of hazard to the skin are known or suspected to be present, and skin contact is possible. • Operations must be conducted in confined, poorly ventilated areas until the absence of conditions requiring Level A protection is determined.	• Fully-encapsulating suit material must be compatible with the substances involved.

B

RECOMMENDED:

- Pressure-demand, full-facepiece SCBA or pressure-demand supplied-air respirator with escape SCBA.

- Chemical-resistant clothing (overalls and long-sleeved jacket; hooded, one- or two-piece chemical splash suit; disposable chemical-resistant one-piece suit).

- Inner and outer chemical-resistant gloves.

- Chemical-resistant safety boots/shoes.

- Hard hat.

- Two-way radio communications.

OPTIONAL:

- Coveralls.

- Disposable boot covers.

- Face shield.

- Long cotton underwear.

The same level of respiratory protection but less skin protection than Level A.

It is the minimum level recommended for initial site entries until the hazards have been further identified.

- The type and atmospheric concentration of substances have been identified and require a high level of respiratory protection, but less skin protection. This involves atmospheres:

  — with IDLH concentrations of specific substances that do not represent a severe skin hazard;

  or

  — that do not meet the criteria for use of air-purifying respirators.

- Atmosphere contains less than 19.5 percent oxygen.

- Presence of incompletely identified vapors or gases is indicated by direct-reading organic vapor detection instrument, but vapors and gases are not suspected of containing high levels of chemicals harmful to skin or capable of being absorbed through the intact skin.

- Use only when the vapor or gases present are not suspected of containing high concentrations of chemicals that are harmful to skin or capable of being absorbed through the intact skin.

- Use only when it is highly unlikely that the work being done will generate either high concentrations of vapors, gases, or particulates or splashes of material that will affect exposed skin.

*Based on EPA protective ensembles.

249

LEVEL OF PROTECTION	EQUIPMENT	PROTECTION PROVIDED	SHOULD BE USED WHEN:	LIMITING CRITERIA
C	RECOMMENDED: • Full-facepiece, air-purifying, canister-equipped respirator. • Chemical-resistant clothing (overalls and long-sleeved jacket; hooded, one- or two-piece chemical splash suit; disposable chemical-resistant one-piece suit). • Inner and outer chemical-resistant gloves. • Chemical-resistant safety boots/shoes. • Hard hat. • Two-way radio communications. OPTIONAL: • Coveralls. • Disposable boot covers. • Face shield. • Escape mask. • Long cotton underwear.	The same level of skin protection as Level B, but a lower level of respiratory protection.	• The atmospheric contaminants, liquid splashes, or other direct contact will not adversely affect any exposed skin. • The types of air contaminants have been identified, concentrations measured, and a canister is available that can remove the contaminant. • All criteria for the use of air-purifying respirators are met.	• Atmospheric concentration of chemicals must not exceed IDLH levels. • The atmosphere must contain at least 19.5 percent oxygen.
D	RECOMMENDED: • Coveralls. • Safety boots/shoes. • Safety glasses or chemical splash goggles. • Hard hat. OPTIONAL: • Gloves. • Escape mask. • Face shield.	No respiratory protection. Minimal skin protection.	• The atmosphere contains no known hazard. • Work functions preclude splashes, immersion, or the potential for unexpected inhalation of or contact with hazardous levels of any chemicals.	• This level should not be worn in the Exclusion Zone. • The atmosphere must contain at least 19.5 percent oxygen.

Inspection is necessary to keep PPE in good working order. PPE should be inspected:

- when received from the supplier
- when issued to workers
- after use in training or in emergency
- periodically while in storage
- when problems arise with similar equipment.

Table 4-10 is a checklist for inspection of PPE.

4.7.2. Safety Showers and Eyewash Stations

Safety showers are necessary in areas where hazardous chemicals are stored, handled, or used. Showers will flood personnel with water and so quickly remove harmful materials, cool the skin, or douse clothing fires. The effects of contact with materials such as acids, caustics, cryogenic fluids, hot fluids, and other hazardous substances can be minimized by the prompt use of safety showers.

Emergency showers should be prominently marked and well lighted. They should be located in conspicuous spots, preferably near normal traffic patterns but not in the vicinity of electrical apparatus or outlets. They should be placed wherever there is a chance of contact with hazardous materials. Specific examples are laboratories and the vicinity of such operations as sampling, drum handling, packaging, and opening of equipment. Showers in remote locations often are equipped with some form of alarm actuation in order to alert central personnel of a possible problem.

The temperature of the water is important. It must not be too hot or too cold because of the risk of further injury or thermal shock. Cold water also is a less effective flushing agent. When it is necessary to heat water, this can be done at a central location from which water is pumped through a loop serving all showers. Alternatively, individual showers can be supplied with heaters. Freeze protection is an important concern, even in some climates where process water systems are not protected.

Response time is critical. This should be a primary emphasis in personnel training, which should cover:

- knowledge of shower locations
- need for rapid entry
- time required under shower
- with certain materials, need to remove PPE and clothing

Proper design and operation of a system require:

- sufficient number of showers
- standardization

TABLE 4-10
Sample PPE Inspection Checklists

CLOTHING

Before Use:
- Determine that the clothing material is correct for the specified task at hand.
- Visually inspect for:
  — imperfect seams
  — nonuniform coatings
  — tears
  — malfunctioning closures
- Hold up to light and check for pinholes.
- Flex product:
  — observe for cracks
  — observe for other signs of deterioration
- If the product has been used previously, inspect inside and out for signs of chemical attack:
  — discoloration
  — swelling
  — stiffness

While in use, periodically inspect for:
- Evidence of chemical attack such as discoloration, swelling, stiffening, and softening. Keep in mind, however, that chemical permeation can occur without any visible effects.
- Closure failure.
- Tears.
- Punctures.
- Seam discontinuities.

GLOVES

•**Before use:** Pressurize glove to check for pinholes. Either blow into glove, then roll gauntlet toward fingers, or inflate glove and hold under water. In either case, no air should escape.

FULLY ENCAPSULATING SUITS

Before use:
- Check the operation of pressure relief valves.
- Inspect the fitting of wrists, ankles, and neck.
- Check faceshield, if so equipped, for:
  — cracks
  — crazing
  — fogginess

SCBA RESPIRATORS

- Inspect SCBAs:
  — before and after each use
  — at least monthly when in storage
  — every time they are cleaned
- Check all connections for tightness.
- Check material conditions for signs of:
  — pliability
  — deterioration
  — distortion

TABLE 4-10 *(Continued)*
Sample PPE Inspection Checklists

SCBA RESPIRATORS (CONTINUED)

- Check for proper setting and operation of regulators and valves (according to manufacturer's recommendations).
- Check operation of alarm(s).
- Check faceshields and lenses for:
  — cracks
  — crazing
  — fogginess

SUPPLIED AIR RESPIRATORS

- Inspect SARs:
  — daily when in use
  — at least monthly when in storage
  — every time they are cleaned
- Inspect air lines prior to each use for cracks, kinks, cuts, frays, and weak areas.
- Check for proper setting and operation of regulators and valves (according to manufacturer's recommendations).
- Check all connections for tightness.
- Check material conditions for signs of:
  — pliability
  — deterioration
  — distortion
- Check faceshields and lenses for:
  — cracks
  — crazing
  — fogginess

AIR-PURIFYING RESPIRATORS

- Inspect air-purifying respirators:
  — before each use to be sure they have been adequately cleaned
  — after each use
  — during cleaning
  — monthly if in storage for emergency use
- Check material conditions for signs of:
  — pliability
  — deterioration
  — distortion
- Examine cartridges or canisters to ensure that:
  — they are the proper type for the intended use
  — the expiration date has not passed
  — they have not been opened or used previously
- Check faceshields and lenses for:
  — cracks
  — crazing
  — fogginess

- easy accessibility
- prominent markings
- positive effort to prevent clutter under or near showers
- easy activation
  —pull bar or chain
  —automatic (activated by stepping onto platform)
- frequent testing
- adequate drains

Safety showers are usually accompanied by eyewash stations. The same considerations apply. In addition, eyewash stations require

- strainers or screens to remove suspended solids
- reasonably low discharge pressure/velocity to prevent damage to the eyes
- more complete and secure absence of harmful contaminants in the water
- enhanced control of temperature

Eyewash stations should be used long enough to flush a material from the eyes. The MSDS may recommend a regimen, and medical assistance should be sought when called for. An injured party may need assistance to keep the eyes open while flushing.

4.8. CONTROL OF WORKING ENVIRONMENT

4.8.1. Temperature Extremes

Personnel exposed to high temperatures, alone or in conjunction with high humidity, can suffer from heat stress. Cold environments with high air velocities can cause frostbite or hypothermia. ASHRAE [11] has established comfort zone standards identifying the temperature and humidity ranges in which individuals can work comfortably and safely. Areas that must operate outside these comfort zones should be

- identified as thermal hazards and designated as limited access areas
- insulated and isolated from normal personnel work areas
- classified as requiring specific personnel training, monitoring, and protective equipment

The tolerance level can be extended when time of exposure is limited. When extra clothing is needed for chemical protection in a warm environment, heat stress becomes more likely.

In a cold environment, plant personnel should be provided with adequate insulative clothing and shelter. Figure 4-10 plots the tolerance of individuals to lower temperatures based on the insulative value (clo) of their protective

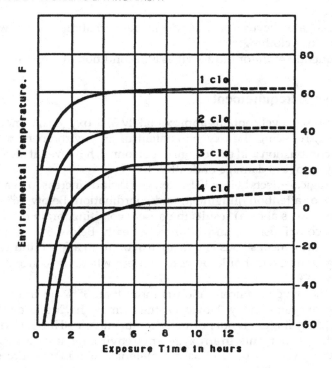

Activity: Light manual work

1 clo = Light coveralls
2 clo = Coveralls + jacket and wool underwear
3 clo = Full intermediate assembly
4 clo = Heavy pile and quilted fabrics

FIGURE 4-10. Tolerance to Cold Environments.

clothing. A practical lower limit for physical activity with the best available insulative clothing is about –30°F.

The following management actions will reduce thermal hazards:

- Carefully train personnel who work in high- or low-temperature environments.
- Carefully schedule work to allow rest periods.
- Provide readily available warm shelters for workers in cold areas and air conditioned shelters for workers in warm areas.
- Allow for frequent replacement of body fluids by workers in warm areas.
- Avoid overprotection. Too much clothing will increase the likelihood of heat stress in a hot environment. Excessive sweating will decrease the insulative value of protective clothing and increase the likelihood of cold-related injuries.

- Investigate the use of a temperature-regulating system within the protective clothing.
- Frequently monitor the activities and condition of workers.

4.8.2. Oxygen Requirements

Normal air at sea level contains approximately 21% oxygen. As mentioned in Section 2.2.1, reducing the oxygen content of an environment reduces the chances of combustion and explosion. In personnel, however, it also can cause serious physiological damage or death.

Physiological effects below 19.5% oxygen include increased heart rate and impairment of attention, judgment, and coordination. Below 16%, hypoxia (also referred to as anoxia) results in nausea, vomiting, brain damage, heart damage, unconsciousness, and ultimately death. Below 6%, an individual rapidly becomes unconscious and death can occur in a few minutes. Oxygen monitoring equipment should be used in areas where the concentration can drop below 20%.

When the oxygen concentration is more than 35%, an individual may suffer from oxygen toxicity. However, the primary hazard is one of fire or explosion. Many materials not normally combustible will burst into flames in oxygen-rich environments. In an oxygen-rich environment, care must be taken to reduce the inventory of combustible materials, eliminate sources of static electricity, and prohibit smoking.

4.8.3. Noise

Excessive noise in an operating plant is not only annoying and distracting but also a hazard. Table 4-11 lists the range of sound intensity and the permissible exposure limits established by OSHA in 29CFR 1910.95. Above 80 dBA, where exposure limits apply, a noise-reduction program is required. Excessive noise can produce several dangerous effects:

- Personnel can be startled, distracted, or annoyed to the point of losing concentration.
- Prolonged exposure can cause pain, physical damage to the ear, and hearing loss.
- Communication is disrupted.

The noise experienced by an individual is a function of

- the number, types, and placement of equipment in his vicinity
- the architectural and structural materials used to mount and house the equipment

Table 4-12 lists a few sources of noise and suggests methods of reduction. Frequently, a source can be placed in a sound-deadening enclosure in order

TABLE 4-11
Permissible Noise Exposure

Sound Level,dBA	Sound Source Example	Permissible Exposure Time, hr
30	Soft whisper	No limit
40		No limit
50		No limit
60	Normal conversation	No limit
70	Loud conversation	No limit
80	Shouting	32
85	Flare stacks (60–90 dBA)	16
90	Boiler room (90–100 dBA)	8
95		4
100	Steam ejectors (90–115 dBA)	2
105		1
110	Thunder	0.5
115		0.25
120 *		
125 *		
130 *	Pneumatic chipping hammer at 3 feet	
140 *	Jet engine	

* Exposures above 115 dBA are not permitted for any duration.

to reduce personnel exposure. Methods for evaluating the combined effect of noise production and noise suppression are defined in 29CFR1910.95.

Where noise cannot be suppressed completely, personnel in the area should be provided with hearing protection equipment. Other personnel should be warned by posing signs around the area of high-intensity noise. Table 4-13 lists four types of protection and their attenuation standards. Test methods are prescribed in ANSI S3.19-1974.

TABLE 4-12
Sources of Noise and Methods of Reduction

Process Equipment or Component	Source of Noise	Suggested Noise Reduction Method
Pumps	Cavitation	Change design
	Solids in liquid	Reduce velocity, if possible
	Worn bearings	Replace bearings
Valves	Choking	Reduce fluid velocity
		Change valve design
	Chattering	Change controls to reduce cycle
Piping	Undersized pipe	Reduce fluid velocity
	Small radius elbows	Change elbow radius
Fans	Stalling	Change gas velocity
		Change fan design
		Change fan wheel design
	Gear reducer	Replace with belts
Compressors	Piping vibration	Install restraints and vibration dampeners
	Antisurge bypass s	Install quiet valve
		Enlarge and streamline piping
Motors	Normal operation	Replace with quieter motor
	TEFC cooling air fan	Install sound enclosure
Boilers & Furnaces	Combustion at burners	Acoustic plenum
		Seal around sight holes and control rods
Ejectors	Normal operation	Install sound enclosure

TABLE 4-13
Attenuation Standards for Hearing Protection Equipment

Frequency, Hertz	Noise Attenuation, dBA			
	Personal ear plugs	Disposable ear plugs	Ear muffs	Ear plugs with plastic band[a]
125	20 ± 7.1	32.6 ± 4.6	14 ± 4.3	18 ± 3.9
250	23 ± 6.0	34.6 ± 4.5	18 ± 4.4	18 ± 5.2
500	24 ± 6.4	36.7 ± 5.4	25 ± 6.3	19 ± 5.7
1000	26 ± 5.7	40.0 ± 3.9	34 ± 7.4	25 ± 5.8
2000	32 ± 4.4	41.2 ± 3.3	39 ± 6.8	35 ± 7.9
3000	41 ± 5.5	46.1 ± 4.5	43 ± 8.0	45 ± 8.2
4000	42 ± 5.3	47.6 ± 4.2	37 ± 7.9	46 ± 7.8
6000	38 ± 6.7	48.1 ± 4.1	26 ± 8.7	46 ± 8.3
8000	35 ± 6.6	43.9 ± 3.8	43 ± 7.9	18 ± 3.9

[a] Values are based on band being worn under the chin. If band is worn

REFERENCES

1. Hull, B., and V. John, *Non-Destructive Testing*, Springer-Verlag, New York, 1988.
2. Sandler, H. J., and E. T. Luckiewicz, *Practical Process Engineering, A Working Approach to Plant Design*, Table 10.4, McGraw-Hill, New York, 1987.
3. Higgins, L. R., ed.-in-ch. *Maintenance Engineering Handbook*, 4th ed., McGraw-Hill, New York, 1988.
4. National Fire Protection Association, *Before the Fire: Fire Prevention Strategies for Storage Occupancies*, NFPA, Quincy, MA, 1988.
5. Cote, A. E., ed. *Fire Protection Handbook*, 16th ed., National Fire Protection Association, Quincy, MA, 1986.
6. Factory Mutual Engineering Corp., *Handbook of Industrial Loss Prevention*, 2d ed., McGraw-Hill, 1967.
7. Bierlein, L. W., *Red Book on Transportation of Hazardous Materials*, Cahners Books International, Boston, 1977.
8. OSHA Notice of Proposed Rulemaking, 54FR24080-5, 5 June 1989.
9. NIOSH Alert, "Request for Assistance in Preventing Occupational Fatalities in Confined Spaces," January 1986.
10. Department of Health and Human Services, NIOSH Publication No. 85-115, *Occupational Safety and Health Guidance Manual for Hazardous Waste Site Activities*, DHHS, GPO, Washington, 1985.

5

CLEANUP AND PROCESS CHANGEOVER

This chapter deals with the problems associated with safe and efficient cleanup operations. It addresses process planning, the cleaning process itself, the problems of changeover, and methods of equipment preparation.

The first section (5.1) deals with three phases of process planning: process development, system design, and production scheduling. The need for cleaning and the safety aspects of the cleaning process should be recognized during each phase.

The second section (5.2) deals with cleanup methods. Cleaning may be mechanical, chemical, or a combination of the two. The hazards encountered in the various methods are quite different, as are the precautions that must be taken. This section will discuss the similarities and differences between the two basic methods.

The last section (5.3) of the chapter deals with safety considerations in changeover. Review of the system design is important to ensure compatibility with new operating conditions and properties of the materials being processed. This review is needed whether we are dealing with a permanent system or one made up of portable, general purpose components. A thorough review of the changeover procedure is crucial to subsequent operation and should include inspection, pressure testing, isolation from dangerous or incompatible materials, and surface preparation.

5.1. PROCESS PLANNING

During process development, cleaning and changeover operations should be incorporated into the overall process system design and schedule. Thorough planning at this stage can reduce the need for cleaning and enhance the safety of operations.

When planning a changeover, procedures must be developed and operations personnel must be made aware of the importance of following these procedures to ensure safe operation. As a minimum, these five actions should be considered:

- Develop and write procedures for changing processes and cleaning equipment.
- Review the safety of these procedures and the materials used for cleaning.
- Validate the procedures.
- Follow the procedures, with no unvalidated shortcuts.
- Monitor the effectiveness of the procedures.

5.1.1. Process Development

Process planning begins in the development stage. One of the results of this phase should be a clear idea of how to clean equipment on a production scale. It is very easy and quite common to put all one's effort into measuring and improving process performance. Proper attention is all too frequently not given to the internal condition of equipment and the limits beyond which fouling or precipitation might occur.

To prevent later problems, specific measurements and observations of fouling conditions should be made in development trials. The outcome should be a comprehensive set of guidelines, checklists, and workable procedures, resulting in the development of a safe, stable, and nonfouling process. Some of the specific items to consider during development are:

- flash points of process materials
- conditions that lead to overreaction (polymerization, charring, etc.)
- sensitivity to impact and shear (agitators, pumps, etc.)
- potential for deposit on heat-transfer surfaces
- localized precipitation due to reaction or over-concentration (e.g., at the ends of dip pipes)
- specific interactions of process materials with both internal and external surfaces
- reactivity between process materials and cleaning reagents or heat-transfer fluids
- effects of process or control upsets
- positioning of isolation valves to minimize dead volumes
- elimination of undrainable segments of process lines
- order of reactant additions to minimize fouling

If cleaning proves to be necessary, possible approaches should be considered during process development. Cleaning methods are broadly classified as mechanical or chemical.

Mechanical cleaning usually requires opening of equipment. Techniques include manual stripping and wiping, rinsing, and blasting with abrasives or high-pressure liquid sprays.

Chemical cleaning is based on reactivity (e.g., oxidation of a waste that fouls equipment) or solvent power. A special category here is the "clean-in-place" (CIP) technique. CIP has the advantage of not requiring that equipment be opened and can save time and minimize exposure to hazards. It also allows the use of more aggressive cleaning conditions and so can result in a higher degree of cleanliness. CIP is discussed in more detail later in this chapter. When cleaning in place is a desirable option, the development phase should define

- composition of the cleaning fluid
- time, temperature, and pressure of exposure
- techniques for recovery or recycle of cleaning fluid
- hazards associated with cleaning chemicals and their reaction with films or deposits in the equipment
- impact of raw material impurities and their effects on safe cleanup and handling
- methods for minimization, recovery, and disposal of wastes

5.1.2. System Design

5.1.2.1. Cleanability

When cleaning needs have been defined, the next step is to determine how they can be met. The design and installation of equipment and piping are critical here.

Proper design can simplify cleaning in two different ways. It can reduce the need for cleaning, for example by injecting a reagent at the optimum location to reduce fouling. The design can also improve the effectiveness of cleaning, for example by making dismantling easier or by locating sprays for best cover of internals. Some considerations in cleaning system design are:

- vessel and batch sizing
- equipment design and materials of construction
- placement of cleanout nozzles
- placement of drain lines and use of flush-bottom valves
- design and location of heat transfer surfaces
- placement of charging points and nozzles
- proper design of internals
- design and layout of line-flush piping
- internal surface finish
- efficient drainage of process piping and equipment

Appropriate selection of equipment size relative to batch size and process sequence can reduce fouling. For example, a reactor or other vessel may have various ingredients added during an operation. It may have part of its contents drained or boiled away and may be cooled or heated. All these changes may result in opportunities to uncover agitator blades or to deposit solids on unswept heat transfer surfaces by precipitation or baking. Process planning should consider such possibilities. By addressing a problem beforehand, the developers and designers should at least be able to mitigate the problem or define the cleaning requirements more precisely.

5.1.2.2. Materials of Construction

The designer must consider also the choice of materials of construction [1–5]. The system must be resistant at cleaning temperatures and pressures to all materials used in the process. These include process ingredients, individually and in combination, reaction products at process temperatures and pressures, and substances used in chemical cleaning.

"Materials of construction" refers to all parts of an apparatus. It is not sufficient to identify something as a "stainless steel vessel" or a "titanium pipe." These materials must be identified by a material code number (ASTM, SAE, etc.) and grade [1]. Nonmetallic components such as gaskets, packing, and seals often require similar detailed identification. Plant personnel must be aware of the nature of these materials and must guard against their replacement with unknown or inferior materials during maintenance. This potential problem should be specifically covered in the management-of-change process.

The chief concern in selection of materials of construction must be the prevention of corrosion and equipment failure. In some cases, minor corrosion may also cause contamination of a product. Such corrosion cannot always be predicted during the design phase and therefore must be studied during process development. A more subtle problem can arise with compound materials. The base material (e.g., a rubber or a plastic) may be inert to corrosion; however, other components of the compound material may cause a problem. The same compounding materials used in product development should therefore be specified for use in production runs. This may be best ensured by using the same brands in all stages of production.

5.1.2.3. Mechanical Design of Equipment

Other important design considerations can be grouped under the heading of mechanical or detailed design of process equipment.

Cleanout Nozzles

Cleanout nozzles have a vital purpose. Unless the equipment can be cleaned properly and thoroughly, a production campaign may fail or a hazard may be created. The nozzles must be sized and positioned properly. Access should be specifically provided to all parts that may require cleaning.

Drains

The location and placement of drainage nozzles are similarly important. It usually is most convenient to have a low-point drain and to design the apparatus for proper drainage to that point. Trays and baffles in vessels may need drain holes. With a vertical dished-headed vessel, a bottom opening at the center line should automatically provide complete drainage of the body of the vessel. Best results and best process design often will call for the use of flush-bottom valves. Piping from a drain must have the proper downward pitch to a collection point and must not contain any risers or upturned loops that inhibit drainage.

Feed Point Location

The position of reactant addition points should be considered. The order and location of addition of key reactants may be critical, especially when proper dispersal is necessary to avoid coagulation or fouling of surfaces. However, if all ingredients are added together, this may be of little importance.

In less sensitive cases, materials may simply be added through a vessel's top head. In other cases, it may be important to introduce a component into a specific region or in a specific form. As examples of the former, favored regions might be beneath a liquid surface, in the vicinity of an agitator or disperser, away from a heated surface, at the bottom of a vessel or column, or parallel to the axis of a centrifuge. As examples of the latter, the physical form of addition may be specified as a liquid spray or discrete droplets. Design should ensure that no important changes will be caused by varying levels in the apparatus, varying delivery pressures, or similar deviations.

Internals

The proper design of internals includes both their functionality and their cleanability. Arrangement of internals can also influence the cleanability of the entire apparatus. Baffles in a vessel are a good example; if improperly arranged, they can interfere with the cleaning process.

Other examples of internals sometimes difficult to clean are coils, dip pipes, instrument probes, agitators, access ports and vessel heads. Nonprocess connections, which are usually dead spaces, can also be troublesome. These connections include pressure and level taps, sample connections, sight glasses, and the like. To avoid cleaning problems, such internals should be evaluated for their influence on the cleaning process during the equipment design phase.

Surface Finish

Another factor of special importance in the design of a cleanable system is the finish on process surfaces [6]. Any process material is retained more readily on a rough surface, simply by accumulating in the irregularities. To avoid this problem, smooth surfaces should be specified wherever possible. For example, stainless steels can be polished to improve their smoothness.

Typical surface finishes and methods for their production are treated in the literature [7]. Special materials of construction should be used if the process material has a strong tendency to adhere to metals (e.g., glass-lined and PTFE-lined steels).

5.1.3. Production Scheduling

Production scheduling is important in multiproduct operations. With proper planning and the ability to hold some product in inventory, campaigns can be extended to reduce the frequency of turnarounds. If the demand for a particular product is small, frequent small batches may best suit the schedule. However, too small a batch in an oversized piece of equipment can result in levels too low for proper agitation or in incomplete coverage of heat transfer surfaces. These conditions may aggravate fouling problems or cause instability in the process.

Similar batch products are often distinguished by minor differences in:

- composition
- solution concentration
- number and amounts of minor additives
- molecular weight or monomer distribution (polymeric materials)
- solvents or raw materials (e.g., toluene and xylenes)

Such differences may often be accommodated without scrupulous cleaning of equipment between batches. If two products are made which differ only in, for example, the amount of antioxidant that is added, only minimal cleaning may be needed before switching production. Certain overhead equipment may not be affected at all.

Consider the following example in which a plant is to make five different products. These involve three different compounds. C_1 and C_2 are quite similar, while D is different in character. Each is produced in one of the two solvents, A and B. The formulations (in arbitrary units) are given in the following example:

PRODUCT	I	II	III	IV	V
Solvent A	100	100	100	100	
Solvent B					100
Compound C_1			20		
Compound C_2				20	20
Compound D	10	30			

There is hardly any difference between Products III and IV, and so one would expect to be able to produce them in succession with a minimum of cleaning. Products I and II also are quite similar, differing only in concentration. These, too, should be produced in immediate succession. If Product II has a substantially higher viscosity, it may be well to schedule it first, since the more thorough cleaning required when ending the production of Compound D will become easier. Product V is the only one based on Solvent B. Its place in the production sequence depends on factors not obvious from our table. While it seems to match Product IV most closely, the entire justification for its existence may be the scrupulous absence of Solvent A. It may be necessary to clean the entire apparatus thoroughly before and after production of Product V.

The most important aspect in planning this production run, therefore, may be to make the processing campaign as long as is practicable in order to reduce the number of times the equipment must be cleaned. The bar chart below shows a typical sequence for the above example.

| II | II | II | Rinse, Drain | I | I | Clean | III | IV | IV | Clean | V | V | Clean |

This example intends only to suggest how the cleaning process might influence production scheduling. Any extra time required for cleaning will reduce the effective capacity of the plant, increase cost, add to the waste disposal problem, and more frequently expose workers to the separate hazard that the cleaning process represents. However, a schedule based solely on the degree of cleaning required between products may not be feasible. Other factors may include toxicity, odor, color, and governmental regulations. Without a clear plan of cleaning requirements, there is no opportunity to balance different production demands.

5.2. CLEANUP METHODS

Cleanup methods may be either chemical or mechanical. Chemical cleaning consists of adding a substance into an apparatus or onto its surfaces which helps to remove any residues of process materials by dissolution or chemical reaction. Mechanical cleaning is the physical removal of such residues, which may be performed manually or with the aid of mechanical devices. Some cleaning processes combine the two methods.

While cleaning methods must be selected on a case-by-case basis, some general guidelines can be applied to specific situations. Some of these are:

Vessels, external
Clean manually with solvents or reagents, in combination with mechanical means if necessary. Waste production should be kept to a minimum. Provide adequate personnel protection and ventilation consistent with hazards associated with cleaning reagents and waste products.

Vessels, internal
Avoid physical entry where possible. Use *in-situ* chemical cleaning. If physical entry is unavoidable, use chemical or mechanical cleaning as required. Waste

production should be kept to a minimum, with adequate vessel drainage. Provide appropriate personnel protection and ventilation consistent with hazards associated with cleaning reagents and waste products.

Process lines, dip pipes, sample taps
Clean in place as much as possible. Replace plugged lines if economically feasible or less hazardous; otherwise, chemical lancing, mechanical cleaning by abrasive blasting, and reaming the lines are possible options. Lines susceptible to plugging should be designed for ease of disassembly and replacement. Provide appropriate personnel protection consistent with hazards associated with cleaning reagents and waste products.

Heat exchanger tubes
Similar to process lines. Replacement is economical only in extreme situations.

Valves
Same as process lines. Some applications require disassembly and blast cleaning in glove boxes. Small valves have also been cleaned by ultrasonic techniques.

Sight glasses
Use chemical cleaning, in situ where possible. Avoid any abrasive mechanical cleaning. Replacement may be the most practical alternative.

5.2.1, Chemical Cleaning

Handling and disposal of cleaning chemicals require the same attention given to process chemicals. Because of the nature of the cleaning process and some of the materials used, hazards may be even greater. Aggressive agents and conditions are used for cleaning; the procedures, equipment and training are therefore as important here as in the actual production operation. Operating documents should give clear and explicit directions on handling of cleaning chemicals and on their application. The Material Safety Data Sheet (MSDS) published for each cleaning agent or compound (see Section 2.1) contains much of the relevant information. The CCPS *Guidelines for Safe Storage and Handling of High-Toxic-Hazard Materials* should be consulted and used as an aid in setting up a safe program.

When establishing a cleaning procedure, plant management must consider

- hazardous properties of the reagent(s), (e.g., toxicity, reactivity, and flammability)
- use of plant personnel or of contract cleaning services (see Section 4.6)
- use of a permanently installed clean-in-place system or a portable system connected to equipment as needed (see below)
- requirements for personal protective equipment for the operation (see Section 4.7)

- compatibility of the cleaning reagent with the process equipment's materials of construction (see Section 3.1)
- potential release of hazardous by-products of the cleaning process

Many different cleaning agents are available for specific uses. The best choice for a given situation depends on

- nature of the contaminant(s) to be removed
- materials of construction of the equipment or system to be cleaned
- method of cleaning to be applied
- hazardous properties of the solvent
- methods available for handling or disposal of the used solvent

Industries such as those dealing with food, pharmaceuticals, or personal care products are regulated by the FDA or the USDA. These agencies may also regulate cleaning compounds and their concentrations.

With all the problems encountered in the handling and disposal of cleaning agents, it is advantageous to minimize their consumption. Several methods may be used to reduce the use of cleaning chemicals.

Reuse and Recycle

If the amount of contamination picked up by the cleaning agent in a single use is very small, it may be feasible to reuse the cleaning solution. Consumption of active agent can be offset by makeup, and accumulation of dissolved impurities can be offset by purging and replacing some of the solution. In a flow-through system, much of the cleaning solution can be recycled, to the same effect and with the same limitations as its reuse.

Mechanical Aids

Contact and sweeping of surfaces by the solvent are necessary for adequate cleaning. These conditions can be achieved by filling a vessel with solvent and then agitating the contents. The same effect can be obtained by using much smaller volumes of cleaning agent in the form of a spray, created by a spray ball or nozzle mounted internally. The spray pattern must be matched carefully with the geometry of the vessel, and more than one spraying device may be needed. Such devices complicate equipment design but may offer large reductions in solvent use when contaminants are easily removed on brief contact with the solvent. However, the use of sprays can result in accumulation of static charges. Section 2.2.5 discusses the hazards of static electricity.

Blending with Product

In some cases, the cleaning solvent may become part of the product. The best example of such blending is a series of products made in the same solvent but requiring some rinsing between batches. If the product can be made at concentrations higher than the sales specification, and if postblending is possible, the rinse would simply be combined with the product solution and become part of the final product.

Multiple Rinses

Multiple rinses with small volumes of solvent will leave equipment cleaner and use solvent more efficiently than a large single rinse.

Cleaning in Place

It is often possible to design and equip a system to virtually eliminate disassembly and opening of lines or equipment during cleaning. The "clean-in-place" (CIP) technique requires facilities for cleaning solution preparation, adjustment of solution to the required temperature, introduction of solution into the apparatus being cleaned, and safe and effective removal and disposal of the spent solution. Figure 5-1 shows a CIP system that may be used for cleaning equipment, in this case a heat exchanger, at remote locations. The system is portable, utilizing a tank truck to carry cleaning agent to the site and a portable or fixed cleaning station that can be hooked up to the heat exchanger.

Many CIP systems require a permanent installation for cleaning solution preparation with temporary hookup to the equipment being cleaned. Minimizing waste production is a vital aspect of the design of any CIP system. Piping and equipment do not need to be completely flooded to be cleaned. The cleaning solution can be injected into the apparatus using a high pressure spray nozzle. Figure 5-2 shows a temporary hookup for cleaning lines with high pressure sprays; Figure 5-3 shows a variety of process vessels containing permanently mounted spray nozzles or spray balls.

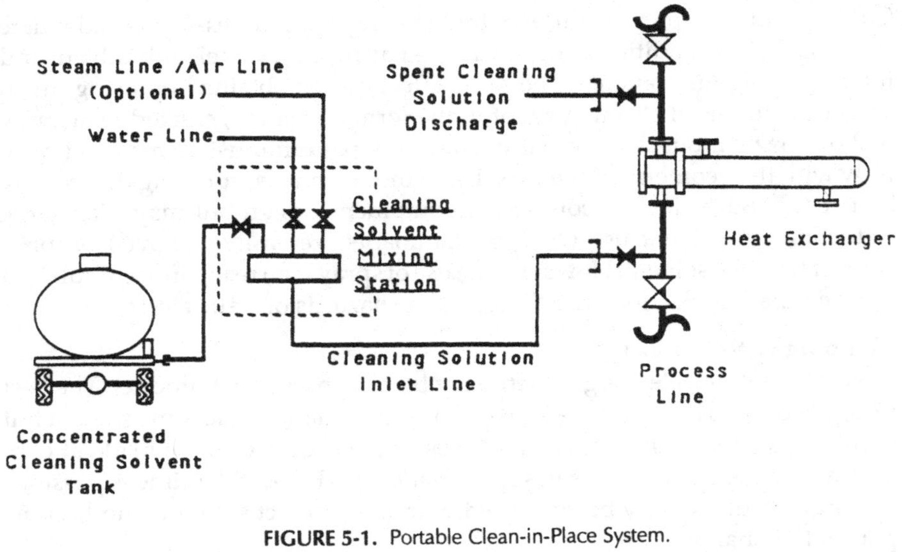

FIGURE 5-1. Portable Clean-in-Place System.

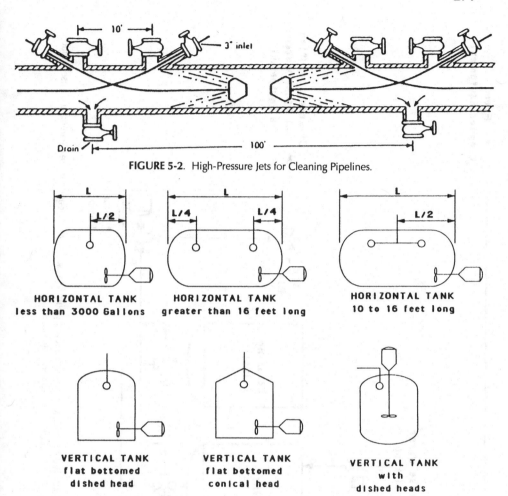

FIGURE 5-2. High-Pressure Jets for Cleaning Pipelines.

HORIZONTAL TANK
less than 3000 Gallons

HORIZONTAL TANK
greater than 16 feet long

HORIZONTAL TANK
10 to 16 feet long

VERTICAL TANK
flat bottomed
dished head

VERTICAL TANK
flat bottomed
conical head

VERTICAL TANK
with
dished heads

FIGURE 5-3. Recommended CIP Spray Nozzle Locations. Note: The spray nozzle should be located 8 to 12 inches below the top inner surface of the vessel.

A typical permanent CIP station is shown in Figure 5-4. A permanent station may simply be a two-tank system, one for cleaning solution and another for rinse water. A third tank allows for collection and recirculation of solution. More tanks can be added depending on the number of cleaning agents and the number of cleaning steps required in the process.

CIP offers a number of advantages:

- improved safety
- reduced exposure of personnel

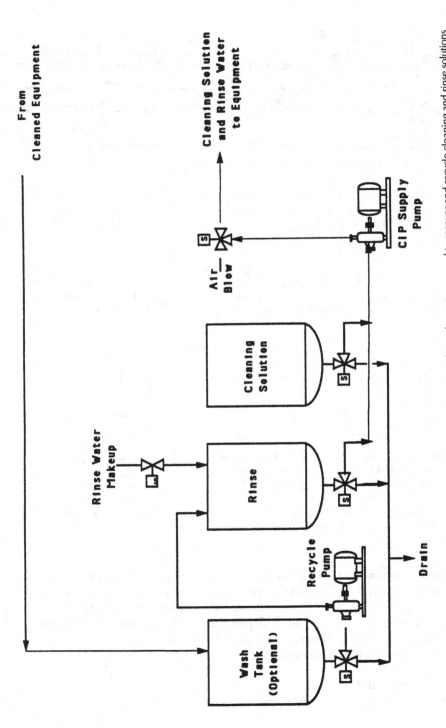

FIGURE 5-4. Typical CIP Equipment Package. Note: The optional wash tank and recycle pump are used to recover and recycle cleaning and rinse solutions in order to reduce waste water.

- consistent results
- labor reduction
- reduction in time required
- reduction of cleaning solution quantity
- possible use of more aggressive conditions or more reactive chemicals
- reduction of mechanical damage during equipment disassembly and reassembly.

The disadvantages of CIP are the cost of the added facilities, the extra space that they require, and the difficulty of retrofitting if not part of the original design.

Before embarking on a CIP design or campaign, the person responsible should establish the following:

- a list, with sizes, of the equipment and pipelines to be cleaned
- identity of the residue and its composition
- difficulty of removal of residue
- definition of "clean" —how much residue can be tolerated, and how can it be measured?
- composition of the cleaning solution
- time and temperature of exposure of the residue to the cleaning process

If cleaning in place is the preferred method, the next decision is whether to use a once-through or a recycle system. The once-through approach is useful when

- the system is easy to clean
- chemical cleaning is infrequent
- temperatures near ambient are adequate
- exposure times are short
- solutions are dilute, of little value, or easily disposed of
- used solutions can be reused or reclaimed

In other cases, a recycle system is more appropriate. However, recycle systems place many more constraints on the design and layout of the total system.

When fixing the details of a CIP system, the first thing to determine is the equipment's conformity with design requirements. These include cleanability and absence of any harmful effects from the cleaning solution. CIP imposes its own demands on system design. Piping should be free-draining and without dead ends. Because some degree of turbulence is necessary to clean a pipeline, a minimum velocity of 5 ft/sec usually is recommended. Many but not all process pumps can be cleaned effectively in place. Some centrifugal pumps are good candidates for CIP, but positive displacement pumps will require dismantling. Tanks may be cleaned by flooding or, as shown in Figure 5-3, by impingement. In either case, proper venting is important. The more elaborate CIP systems provide separate tanks for preparation of cleaning solutions and allow gravity drainage to used-solution tanks.

Vigorous cleaning may require the use of chemicals at high temperature or high pressure. These conditions present an additional hazard which may call for temporary protection (e.g., personnel barriers, flange shields), particularly in systems that are not well insulated or are not operated regularly under pressure. The nature of the cleaning agent may also call for special measures. Special solvents, acids, oxidizers, and strong alkalis are frequently used in CIP systems. Each chemical must be handled with all precautions identified by the supplier or on the MSDS. These precautions will often be different from those taken in routine operation and will require extra safety gear. Those involved in the cleaning operation should work from a checklist that details all these requirements.

Another chemical cleaning hazard is reaction between the cleaning solution and the residue. Some agents are strong oxidants; other may cause decomposition of the residue. These reactions may also evolve heat or produce gaseous by-products that must be vented properly. If the cleaning by-products are hazardous, they may require destruction or capture and disposal. Heat of reaction may cause boiling or generate pressure and force the cleaning solution to escape or flow into other sections of the apparatus.

Finally, the reaction may increase corrosivity of the solution. Many of these problems can be anticipated from the nature of the cleaning agent and the process conditions used during cleaning. Other potential problems may be more specific to the nature of the plant's products and will require individual attention. The best time to identify these problems and develop safe solutions is during process development as described in Section 5.1.1.

Boilout
Systems frequently are cleaned by applying heat and boiling a solvent or cleaning solution in a reactor or other vessel. The hot solution can also be circulated to other parts of the system, and accessible parts of an overhead system can be cleaned by reflux from a condenser. Many such systems have provisions for temporary connections. In all cases, provisions for cleaning must be included during design.

A frequent hazard is the accumulation of high-boiling or unstable compounds. These may be routinely present in the process or may form by the reaction of residues when the process is off line. A good example are peroxides, which may be concentrated by boiling. The input of heat may cause their rapid, even explosive, decomposition. Peroxides can also generate high oxygen levels, which can create fire or explosion hazards if organics are present. Small quantities can dry and decompose when thin films of solution are exposed to heated surfaces. A boilout process must be controlled carefully to avoid this hazard. In all cases, unstable compounds must not be allowed to reach dangerously high concentrations by continued reaction or by evaporation of some lower-boiling diluent.

5.2.2. Mechanical Cleaning

Mechanical cleaning involves the physical application of a force to clean a surface. Different types of force can be used in combination to achieve the required results. Depending on the type of applied force (manual, motive, or vibratory), mechanical cleaning can be classified as manual, abrasive, or ultrasonic. Each of these is discussed below.

Unlike chemical cleaning, most mechanical cleaning techniques involve some degree of surface removal. Those processes which produce cutting and gouging are more hazardous than those which produce polishing. The process facility should have a set of guidelines as to when mechanical cleaning should be used. The following should be considered when establishing the guidelines:

- equipment disassembly or cleaning in place
- identification of surfaces to be cleaned (internal and external)
- cleaning equipment to be used
- cleaning solutions and reagents
- protective equipment to be worn
- communication
- posting, marking, or isolation of area

5.2.2.1. Manual Cleaning
Manual cleaning is one of the most hazardous, labor intensive, inefficient, and expensive methods for cleaning process facilities. It is limited to surfaces that are relatively easy to reach. Chemical cleaning should be used to remove unwanted materials and contaminants in crevices, difficult-to-reach surfaces, and remote corners.

5.2.2.2. Abrasive Cleaning
With this method, an abrasive material is applied by a fluid at high velocity against the surface to be cleaned. Depending on the fluid used, a system is classified as dry or wet.

Harder, less friable materials clean faster and are more economical than softer, friable abrasives. However, hard materials produce more cutting and gouging. If surface depletion due to cutting and gouging is unacceptable, a softer material must be used, at the expense of speed.

Typical materials used as hard abrasives are sand, silicon carbide, bits of steel wire, steel shot, iron grit, glass beads, washed silica, silica flour, aluminum oxide and garnet. Some softer abrasive materials are cracked nut shells, plumstones, rice hulls and ground corn cobs. The softer abrasives are good for cleaning and polishing but are not as good for scouring.

Dry Systems
Dry blast cleaning is a relatively simple system that uses compressed air to provide the motive force to the abrasive material. Air pressure for dry systems

ranges from 10 to 120 psig. Air at 90 psig and above results in a faster, more efficient operation but causes more erosion of surfaces. Cleaning at pressures below 60 psig is slower but causes less surface attrition. The high-pressure air is one of the principal hazards in dry blasting. The severity of the hazard increases with the pressure of operation and the volume of the system. Temporary connections are common. Makeshift facilities must be avoided, the connecting lines must be kept in good repair, and the area should be posted or isolated.

Blast cleaning produces a significant amount of dusting. The amount of dust depends on the friability of the abrasive material. Dust particles can settle in mechanical equipment as well as in the clothing of personnel. Dust control procedures should therefore be established as part of the abrasive blasting program.

Wet Systems
Hydroblast or aquablast cleaning systems use a very high pressure water jet as the abrasive or water as the motive fluid with a suspension of abrasives. Wet blasting systems allow the use of finer abrasives and produce finer finishes after cleaning. Finer abrasives cause less surface wear. Surface losses of one mil or less can be achieved. However, wet blasting can also be set up for rapid, coarse cutting. Another advantage of wet systems is the lack of dust production. A disadvantage, however, is a waste liquid stream that must be treated and disposed of. A wet blasting system must be properly designed and operated to minimize splashing and waste production.

High pressure is again a hazard. While the rapid release of large volumes of fluid, with production of shrapnel, is not so much of a hazard as in the case of dry blasting, protection against any hazardous properties of the liquid is necessary.

5.2.2.3. Ultrasonic Cleaning
Ultrasonic cleaning is based on a process called cavitation, which acts on a surface like tiny scrub brushes. It is particularly useful for cleaning disassembled and partially disassembled process components of intricate geometry that would otherwise be difficult to clean. Ultrasonic cleaning can break up or remove unwanted materials in cracks and crevices. A typical ultrasonic cleaning unit comprises an ultrasonic generator, a transducer, and a tank containing cleaning reagent.

The following items should be considered in establishing ultrasonic cleaning procedures:

- Consult equipment vendor data to ensure that the system is properly designed and installed. This includes:
  —correct transducer size and design
  —correct transducer positioning
  —multiple transducers operating in phase to avoid disruption of the sonic waves

- Make sure the tank is the proper size.
- Make sure the tank is clean.
- Determine the nature of the contaminant on the component to be cleaned and establish the following:
  —number of reagents needed
  —with multiple reagents, a proper sequential cleaning procedure
  —advantages of elevated temperatures (operating temperatures must not exceed the Curie point of the transducer)
- Make sure that any cleaning reagent used is compatible with the materials of construction of the cleaning tank and the transducer.
- When immersing the piece to be cleaned in the tank, determine if
  —the piece can be placed as-is into the tank
  —it should be supported or placed in a wire basket
  —the piece should be rotated to expose all surface areas to cavitation
  —a sonic probe should be used to clean the hard-to-reach places

5.2.2.4. Hazards of Mechanical Cleaning

Because mechanical cleaning is labor-intensive, it can expose personnel to hazards not experienced in normal operation. Some of these hazards are presented below.

Manual Cleaning Hazards
- confined spaces with poor ventilation and limited scope for positioning cleaning equipment
- handling of cleaning equipment
- exposure to cleaning solutions or waste

Blast Cleaning Hazards [8]
- confined spaces with poor ventilation and limited scope for positioning cleaning equipment
- respiratory hazards from dust, flying debris, or aerosols
- reagents, dust and aerosols may be deposited on workers' clothing and cause epidermal problems
- poor ventilation for wet blast cleaning may create respiratory hazards from splashing reagent or flying debris
- above hazards can be enhanced when dealing with biological or radiological contaminants
- high pressure and force can cause severe bodily injuries

Ultrasonic Cleaning Hazards
- Immersion of any part of the body in the cleaning solution of an operating ultrasonic cleaner will result in a tingling sensation and pain from the cavitating solution. Prolonged exposure will cause tissue damage.

- Although ultrasonic noise is not audible, exposure to high intensities in the 20 to 25 kHz range may be hazardous. The recommended practice is to limit noise intensity to 75 dB at 20 kHz and 110 dB at 25 kHz and above.
- Debris aerosols are produced from cavitation in the cleaning bath. These aerosols can be extremely hazardous when dealing with biological or radiological contaminants.

5.2.3. Direct Entry

Visual assurance of cleanliness is best achieved by direct entry and hand cleaning, by one means or another, of the interior of equipment. However, vessel entry is one of the most hazardous activities in the chemical industry and should be avoided whenever possible. Potential problems include:

- oxygen deprivation
- hidden chemical hazards
- insecure footing
- mechanical hazards of internals
- awkward body positions

A vessel large enough for entry will almost certainly come under OSHA's definition of a "confined space." Regulations and permits for such work are covered in Sections 4.5.1 and 7.1.4.

Vessel entry may appear to be a simple and direct approach to a stubborn cleaning problem. However, in light of the hazards noted above, it should not be attempted without careful planning and supervision. A generalized check list should include

- consideration of residual chemical hazards and methods of reduction before entry
- locking out of any electrically driven equipment, such as agitators
- positive protection against entry of any unwanted material (e.g., by breaking connections)
- careful planning of work in restricted spaces
- methods for removal of hazardous atmospheres and provision of an adequate air supply
- testing of the vessel atmosphere for oxygen and hazardous vapors before entry
- continuous presence of supervision
- use of all required protective equipment

5.3. PROCESS CHANGEOVER

This section includes design considerations that should be reviewed before an existing system is modified or a new product manufactured. Much of the material included here also applies to a new facility or to the checks that should be made occasionally even in the absence of a changeover.

Subjects covered include materials of construction, sealing devices and their auxiliaries, equipment design limitations, relief system capabilities, interlocks, and testing for removal of chemicals between processes. The objective of these guidelines is to allow review of design of a process system before introducing a new process.

5.3.1. Process Equipment Compatibility

Changeover Concerns

A change in process or formulation demands a review of system components. This review should include, at a minimum, the items listed below.

- electrical classification
- interlocks
- equipment and vessel design
- relief systems
- emission control
- waste handling
- materials of construction
- gaskets
- seals
- packing
- cleaning solutions
- seal fluids
- inerting media
- heat-transfer fluids

5.3.2. Equipment Preparation

Topics in this section include pressure testing to ensure that assembly meets the varying requirements of the process, special methods for thorough removal of traces of vapors, the importance of surface passivation in preventing corrosion, and the avoidance of accidental mixing of the wrong materials.

5.3.2.1. Pressure and Leak Testing

Many different fluids are used in the pressure testing of piping and equipment. The most common are water, air, inert gases, and process fluids. Each option is discussed below.

Some general considerations in preparation for testing are:

- The system chosen for testing should be as large as can conveniently be handled in one test (see below for limitations in high-pressure gas testing).
- All piping components that had been removed should be reinstalled.
- All valves in the system, other than vents and drains, should be open.
- All rotating equipment should be removed from the system.
- All large tanks and vessels should be physically disconnected or blinded (all blinds should have visible tags).
- All branch instruments, relief devices, etc., should be removed.
- All instrument connection lines should be isolated and tested separately.
- All joints, including welds and laminations, should be visible and not covered (e.g., by insulation).
- Working fluid impurities should be known and should be compatible with process fluids and materials of construction.

Water is often the most convenient testing medium, but several reasons may preclude its use. These include

- possibility of freezing
- proper grade of water not available in sufficient quantity at reasonable cost
- need to keep apparatus dry (e.g., incompatibility of water with process materials)
- possibility of corrosion by water not removed after test

Equipment test pressures are often specified in the design documentation. Piping test pressures can be set by discussion with the plant designer or piping supplier. At least two independent pressure gauges of known accuracy should be used. Before beginning the pressure test for equipment integrity, it may be useful to apply a low pressure test with compressed air to check for faulty connections.

When using a liquid, the testers should fill the apparatus from a low point while venting carefully. After filling the system, they should increase the pressure in stages or at a controlled rate consistent with equipment design. The designated test pressure should never be exceeded. The pressure should be released immediately when a leak is detected, vacuum relief must be provided during draining, and the apparatus should not be allowed to remain full of water when there is any danger of freezing. Any components that can interfere with drainage should be removed and tested separately. Pipelines or vessels that normally carry gas may need temporary additional support while full of liquid.

The use of gases for pressure testing increases the hazard in case of equipment failure. The increase in gas volume upon release is a hazard in itself.

There is also the possibility of injury by shrapnel. Normal practice with gases, therefore, is to increase the pressure in stages, allowing ten or fifteen minutes at each level for observation. The volume of the system to be tested often is restricted as a function of the ultimate test pressure. It becomes especially important to arrange for removal of pressure if problems develop.

Process fluids can be used for testing only with discretion. In no case should they be used at high pressure if flammable, toxic, or otherwise injurious to the body. An alternative sometimes used when aqueous testing is impractical and process fluids are hazardous is testing with another fluid that is less hazardous.

Other general safety precautions, regardless of the testing medium, are:

- no mechanical loads on apparatus
- no rapid temperature changes
- work area roped off and restricted to qualified, essential personnel
- no other work in area during test
- no repair work while process material is present in the apparatus
- no overstressing of bolts in attempted repairs

The above discussion is based on the use of metallic pipe and equipment. Because of its slow expansion while under pressure, plastic piping requires stepwise application of pressure and more proof time at the maximum pressure. This procedure is a function of the size and material of the piping and can be set by consultation with the supplier.

5.3.2.2. Special Drying Requirements
In ordinary drying, the presence of residual vapor from test or cleaning fluids is not a detriment if the temperature is high enough. Water can be removed, for example, by blowing steam through a system, as long as there is no condensation.

This process will leave traces of adsorbed water on surfaces, and there may be considerable amounts of water held by gaskets, packing and the like. When requirements are more stringent, steam drying may have to be followed or replaced by a more thorough method, such as hot air drying.

5.3.2.3. Surface Passivation
Often it is not the base material of a metal that resists corrosion. Certain modifications to a surface may render it less reactive; this process is known as passivation.

One such example is titanium, which is in fact a rather reactive metal. Its wide use as a corrosion-resistant material depends on the fact that its surface is easily passivated by oxidation. Most high-level applications of titanium, therefore, are in highly oxidizing media. It is the metal of choice for wet chlorine, for example.

Nickel requires a different sort of passivation protection. Nickel can be oxidized to fairly soluble or unstable hydroxides. It is, therefore, protected by addition of hydrogen donors which can reverse this oxidation reaction.

Passivation is a highly complex subject that cannot be covered effectively within the scope of this book. It is important that the operator be made aware that surface passivation may be critical to safe performance. This awareness is especially important when changing processes and after cleaning surfaces, because the cleaning process often destroys the passivating layer.

5.3.2.4. Incompatible Materials

A problem associated especially with multipurpose plants is the possible entry into an operating system of chemical materials that are extraneous to the process in operation. For example, a solvent or raw material may be awaiting transfer to later use in another process in the same equipment. If incompatible with any of the materials used or produced in the process on line, the reserved solvent/raw material is a potential hazard unless positive steps are taken to exclude it from the ongoing process. Accidental intermingling is a two-way hazard. It is equally possible that a substance can leak from the process into the extraneous solvent/raw material. Inadvertent mixing can also cause difficulties in maintaining product quality and can present extra hazards to customers.

A useful practice, therefore, is to remove the possibility of such accidental mixing. This can be done by physically disconnecting transfer lines or inserting blinds into them. Inadvertent pumped transfers can be avoided by disconnecting or locking out motor starters.

It may happen that two materials that should not be directly mixed are used at different stages of the same batch or are intended to be mixed only when properly diluted by other components of a batch. It is necessary to keep such materials entirely separate outside the equipment in which they are to be mixed. This includes avoiding any cross-connections between their feed systems and ensuring that the fluids are led to separate nozzles.

REFERENCES

1. Sandler, H. J., and E. T. Luckiewicz, *Practical Process Engineering, a working approach to plant design,* Chapter 5, Material Selection, McGraw-Hill, New York, 1987.
2. Budinski, K. G., *Engineering Materials: Properties and Selection,* Reston Publishing Company, Reston, VA, 1978.
3. McNaughton, K. J., *Materials Engineering I—Selecting Materials for Process Equipment,* Chemical Engineering Magazine, McGraw-Hill, New York, 1980.
4. Ibid., *Materials Engineering II—Controlling Corrosion in Process Equipment,* Chemical Engineering Magazine, McGraw-Hill, Inc., New York, 1980.

5. Schweitzer, P. A., *Handbook of Corrosion Resistant Piping*, The Industrial Press, New York, 1969.

6. Villafranca, J., and E. Monroy Zambrano, "Optimization of cleanability," Pharm. Eng. *5*, No. 6, 28, 1985.

7. Ayres, J. A., ed., *Decontamination of Nuclear Reactors and Equipment*, The Ronald Press, New York, 1970.

8. Sanders, M. S., and E. J . McCormick, *Human Factors in Engineering and Design*, 6th ed., p. 472, McGraw-Hill, New York, 1987.

6
TRAINING

This chapter explains the need for safety training at all levels of an organization. Management must develop and require the use of a training plan, with clear directions for implementation, evaluation, and documentation.

The following sections spell out some of the legal requirements for training (6.1), identify those persons who require safety training (6.2), and present some of the methods which can be used (6.3). The extent and frequency of training are also discussed (6.4; 6.5). These will depend on each individual's job function and experience. Finally, the chapter gives a brief summary of record-keeping goals (6.6) and an extensive listing of available training materials (6.7).

6.1. REASONS FOR TRAINING

Training is intended to ensure that employees perform their jobs properly, thus promoting safety and increasing productivity. Many regulatory agencies now specify minimum training requirements. OSHA regulations affecting the process industry are contained in 29CFR1910 and 1926. The Department of Labor publication OSHA 2254, "Training Requirements in OSHA Standards and Training Guidelines," gives a detailed list of OSHA requirements. An important aspect always is the testing of participants to demonstrate that they have understood the training.

6.2. PERSONNEL REQUIRING TRAINING

The following groups should receive appropriate safety training:

- managers and technical personnel
- health, safety and security professionals
- supervisors
- process plant operators

- housekeeping/warehouse personnel
- maintenance and skilled craftsmen; e.g., electricians, crane operators, truck and mobile plant drivers, and welders
- workers with special duties in emergencies; e.g., part-time firemen, first-aid technicians, and telephone operators
- outside contractors (see Section 4.6)

Extent of training is the subject of Section 6.4, which presents specific recommendations on course material for new employees, transferred employees, process plant operators, and supervisors.

6.3. TRAINING METHODS

The method selected should

- match the training to the employee's function
- allow adequate time
- balance classroom and "hands-on" training for best results
- provide feedback from trainees
- allow measurement of effectiveness (has the employee understood the training?)

6.3.1. On-the-Job Training

On-the-job training (OJT) is popular because its hands-on approach makes it attractive for the participant. This method of instruction includes the following categories:

- Job safety analysis (JSA)
- Job instruction training (JIT)
- Over-the-shoulder coaching.

6.3.1.1. Job Safety Analysis (JSA)
The JSA technique identifies hazards in each step of a job and attempts to eliminate them by specifying safe procedures or modifying existing procedures. The analysis is usually recorded on a three-column form similar to that provided by the National Safety Council and shown in Figure 6-1.

6.3.1.2. Job Instruction Training (JIT)
Job instruction training (including over-the-shoulder-coaching), teams an experienced employee with a trainee who is taught each job skill from a prepared schedule or training manual. The degree of difficulty and the intensity of the training can be adjusted to the employee's experience and level of understanding. Job instruction training comprises four parts:

JOB SAFETY ANALYSIS TRAINING GUIDE	Job:		Date:
	Title of Person who Does Job:	Foreman/Supv:	Analysis by:
Department:	Section:		Reviewed by:
Required and/or Recommended Personal Protective Equipment:			Approved by:

SEQUENCE OF BASIC JOB STEPS	POTENTIAL ACCIDENTS OR HAZARDS	RECOMMENDED SAFE JOB PROCEDURE
Break the job down into its basic steps, e.g., what is done first, what is done next and so on. You can do this by 1) observing the job, 2) discussing it with the operator, 3) drawing on your knowledge of the job, or 4) a combination of the three. Record the job steps in their normal order of occurrence. Describe what is done, not the details of how it is done. Usually a few words are sufficient for each step. For example, the first basic job step in using a pressurized water fire extinguisher would be: 1) Remove the extinguisher from the wall bracket.	For each job step, ask yourself what accidents could happen to the person doing the job step. You can get the answers by 1) observing the job, 2) discussing it with the operator, 3) recalling past accidents, or 4) a combination of the three. Ask yourself: can he be struck by or contacted by anything; can he strike against or come in contact with anything; can he be caught in, on, or between anything; can he fall; can he overexert; is he exposed to anything injurious such as gas, radiation, welding rays, etc.?	For each potential accident or hazard, ask yourself how to avoid the accident. You can get your answers by 1) observing the job for leads, 2) discussing precautions with experienced job operators, 3) drawing on your experience, or 4) a combination of the three. Be sure to describe actions specifically. Don't leave out important details. Number each recommended precaution with the same number you gave the potential accident (see center column) that the precaution seeks to avoid. Use simple do or don't statements such as "Lift with your legs, not your back." Avoid such generalities as "Be careful", "Be alert", "Take caution", etc.

FIGURE 6-1. Job Safety Analysis Form [1]

287

- preparation
- presentation
- application
- testing

The trainee's instructor must analyze the job and check the following:

- adequacy of each step in the standard procedure being taught
- health risks and countermeasures
- safety of methods used for handling materials
- opportunities for errors and their consequences
- equipment hazards and provisions against them
- protection against fire and explosion
- potential emergencies and their control
- emergency shutdown procedures

6.3.2. Group Training Techniques

Group training techniques encourage audience participation and a free exchange of ideas and information. This method of instruction can involve any combination of several forms:

- *Conference Method*—the instructor controls the session as the participants share their knowledge and experiences.
- *Brainstorming*—used to find new, innovative approaches to safety-related issues.
- *Case Study*—actual or fictitious cases are used to develop insight and problem solving skills of group members.
- *Discussion*—used for the free exchange of ideas among a small group of individuals.
- *Role Playing*—particularly useful in identifying and changing personnel related to safety.
- *Lecture*—used to impart safety information to a large group in a relatively short time.
- *Question and Answer Periods*—helpful in clarifying a variety of safety issues.

Further discussion of these various methods can be found in some of the material listed in Section 6.7.

6.3.3. Individual Techniques

Many different techniques are available for training individuals. In addition to on-the-job training (Section 6.3.1), techniques include drill, demonstration, quizzes, video, reading, and simulation.

- Drill is normally used to develop worker skill in fundamental tasks.
- Demonstration affords first-hand experiences.
- Quizzes are used to assess the individual's grasp of the subject matter.
- Video and reading techniques are used to enhance instructor-presented materials.
- Simulators are used when it is not feasible to use actual materials or machines.

There are available today many interactive computer training programs. An individual has some control over the time and place of training. The information presented to different individuals is consistent, and on-line testing helps to insure that one topic is understood before a more advanced topic is presented.

6.3.4. School Courses

School courses can take the form of independent study, seminars or short courses. The low-cost method of home or office study uses textbook assignments followed by self-tests. Seminars and short courses on safety are offered by many colleges and private organizations. Section 6.7.1 also lists seminars prepared by safety-oriented groups such as the AIChE.

6.4. EXTENT OF TRAINING

Organizational policy should define training programs for all levels of management and employees. Objectives should be established and then reviewed annually. The following are essential features of a training program:

- All new employees should receive indoctrination training which includes relevant health and safety information.
- All new or reassigned employees should receive job-specific instructions, including relevant health and safety requirements.
- All employees who will participate in mandated PSM programs should be trained in the techniques which will be used (see Section 7.1.1).
- All training programs should specify
  —objectives
  —name(s) of the person(s) responsible for the program
  —trainees who will participate
  —time and place of training
- All training programs should
  —build upon the trainee's existing skills, knowledge and experience
  —define the standard of performance required after training
- Trainee job performance should be monitored during and after training.

- All training should be recorded to show
  —names of participants
  —dates
  —training yet to be completed
  —time taken to reach acceptable performance
- All training programs should be monitored and revised as necessary [2].

6.4.1. Lesson Plans

A methodical and well-planned approach will improve the quality of safety training. One method to ensure thoroughness and efficiency is the preparation of a lesson plan. The lesson plan is designed to help the instructor

- present material in the proper order
- emphasize material according to its importance
- avoid omission of essential material
- keep classes on schedule
- stimulate trainee participation
- increase trainees' confidence [3]

The lesson plan should cover

- identification of the subject matter
- objectives
- training aids
- introduction that establishes the scope of the subject
- method of teaching to be used; e.g. lecture, discussion, etc.
- indication of how the information will be used by the students
- summary of key points
- tests as applicable
- assignments for future classes
- type of certificate to be given

6.4.2. Course Outlines

Section 6.2 identified those individuals who should be given safety training. Recommendations for the scope of the training course for several key groups are given below.

6.4.2.1. New/Transferred Employees
New employees as a group will not be familiar with company operations and policies, nor with the types of hazard they may encounter. Their training program must be intensive in basic material and should include

- company policy statements
- safety and health policy statement (if separate)

- hazard communication
- housekeeping standards
- personal protective equipment
- emergency response procedures
- accident reporting procedures
- near-miss incident reporting
- accident investigation (supervisors)
- lockout/tagout procedures
- machine guarding
- electrical safety awareness
- ladder use and storage
- confined-space entry
- hot work permit procedures
- medical facility support
- first aid/CPR
- hand tool safety
- ergonomic principles
- eye wash and shower locations
- fire prevention and protection
- access to exposure and medical records [4]

Transferred employees may be well versed in company policies and general procedures. They should not, however, be assumed to be well informed about specific hazards and safety procedures in their new jobs. Relevant parts of the new employee training program should be available to all transferred employees.

6.4.2.2. Process Plant Operators
Operators' mistakes can have disastrous safety consequences. Because high levels of skill are required in many operating jobs, proper training is essential. A training course for process plant operators should include

- process goals, constraints, and priorities
- process flow diagrams
- process hazards
- unit operations
- process reactions, thermal effects
- control systems
- process materials, quality, yield
- process effluents and wastes
- plant equipment
- instrumentation
- equipment identification
- equipment manipulation
- operating procedures

- critical process safety parameters
- critical safety devices/controls
- equipment maintenance and cleaning
- use of tools
- permit systems
- failure of equipment and services
- fault administration
- emergency procedures
- fire fighting
- communications, record-keeping and reporting [5]

6.4.2.3. Supervisors

Supervisors play a critical role in ensuring that company safety policies are observed. Their many duties associated with safety may include

- establishing safe operating methods
- instructing workers in the performance of their jobs
- assigning people to tasks
- supervising people at work
- maintaining equipment and the workplace
- maintaining adequate safety equipment and supplies
- maintaining discipline

Supervisors' previous training will have included courses discussed in Sections 6.4.2.1 and 6.4.2.2. Additional courses should supplement the previous training and cover the following:

- communications
- employee involvement in safety
- safety training
- industrial hygiene and noise control
- accident investigation
- safety inspections
- materials handling and storage
- machine safeguarding
- electrical safety
- fire safety
- permit systems

6.5. FREQUENCY OF TRAINING

Training must be frequent enough to communicate and reinforce all required information. Since people learn at different rates, it is difficult to set hard and fast rules for the frequency of training. Some guidelines, however, have been established. A safety training/retraining program is required whenever:

- new employees are hired
- current employees are assigned new tasks
- new equipment or processes are introduced
- procedures are changed
- new information is available or required
- employee performance needs improvement
- interest in safety and efficiency needs a boost [6].

Some indicators of the need for additional training include:

- relatively high accident or injury rates for the industry or the type of work performed
- increasing accident rates
- high employee turnover
- excessive waste or scrap from operations
- company expansion or procedural changes
- changing regulatory requirements
- poor job satisfaction or morale [7]

6.6. RECORD KEEPING

With the emphasis recently placed on safety training, a corresponding need has developed to maintain accurate records. Records should achieve the following:

- ensure that every member of the organization has received the appropriate training and is scheduled to receive additional instruction when required.
- help identify the strengths and weaknesses of the safety program.
- fulfill legal obligations.

Safety records can contain a variety of information, but at the minimum, the following should be noted for each program and each employee:

- course title
- course objectives
- training methods used
- training aids or course materials
- instructor qualifications
- method of evaluation and measured results of training
- dates and duration of training
- training yet to be completed
- time taken to reach acceptable performance

6.7. SOURCES OF TRAINING INFORMATION

Many of the technical societies and trade associations listed in Section 7.3.2 and 7.3.3 publish information on safety and accident prevention. This following section is a partial listing of such offerings, divided into five categories:

- Seminars
- Printed material (pamphlets, booklets, papers and books)
- Audiovisual material (including video tape, slides and film)
- Training kits
- Computer software

Several AIChE courses appear in the first category, as examples. Many other courses are available from time to time, and the list is evolving.

6.7.1. Seminars

Title	Organization
Safety in Pilot Plants and Research Operations	American Institute of Chemical Engineers (AIChE)
The New OSHA Regulation: Process Safety Management of HighlyHazardous Chemicals—29 CFR 1910.119 How to Implement It.	AIChE
Fundamentals of Chemical Process Safety	AIChE
Fundamentals of Fire and Explosion Hazards Evaluation.	AIChE
Chemical Plant Accidents: A Workshop on Causes and Preventions. The Concerns of OSHA 1910.119	AIChE
Industrial Toxicology	AIChE
Process Safety Management: Design and Evaluation of Process Safety Management Systems	AIChE
Prevention and Mitigation Techniques for High Toxic Hazard Releases	AIChE
Use of Hazard & Operability Studies in Process Risk Management	AIChE
Industrial Process Safety Applications	American Society of Safety Engineers (ASSE)

Corporate Safety Management	ASSE
Safety Management Part I: Fundamental Concepts	ASSE
Safety Management Part II: Program Management and Evaluation	ASSE
Occupational & Environmental Safety Part I: Legislation, Regulations & Standards	ASSE
Occupational & Environmental Safety Part II: Applications & Techniques	ASSE
Confined Space Entry & OSHA Requirements Including Lockout/Tagout	ASSE
Confined Space Rescue	ASSE
Electrical Hazards, OSHA and the National Electrical Code, Including Lockout/Tagout	ASSE
Construction Industry Safety & Health Standards	ASSE
Chemical Process Safety Management	ASSE
Trenching and Shoring	ASSE
Overview of Present and FutureEnvironmental Regulations	ASSE
Fire Protection and Means of Egress	ASSE
Providing Physical Protection: Special Hazards	Factory Mutual Engineering Organization (FMEO)
Effective Training Skills	FMEO
Warehouse Hazards & Protection	FMEO
Avoid Disaster by Design	The Center for Professional Advancement (CPA)
Loss Prevention in the Process Industries	CPA
Process Risk Management	CPA
Regulatory Affairs Management in the Chemical and Allied Industries	CPA
Grounding and Protection of Industrial and Utility Distribution Systems	CPA

Chemical and Safety Management	CPA
Environmental Risk Assessment of Chemicals	CPA
Evaluation and Control of Process Hazards	CPA
Explosions and Related Process Hazards	CPA
Hazardous Materials Crisis Management	CPA
Industrial Hygiene Practice	CPA
Industrial Loss Prevention and Risk Assessment	CPA
Quantitative Risk Assessment in the Chemical Process and Related Industries	CPA
Respiratory Protection	CPA
Safe Handling of Compressed Gases	CPA
Safety in the Chemical Laboratory	CPA
Safety in the Chemical Pilot Plant	CPA

6.7.2 Printed Material

Title	*Author*	*Organization*
Handbook of Rigging	W. E. Rossnagel	American Society of Safety Engineers (ASSE)
Trench Safety Shoring Manual: Using Common Sense in the Common Trench	R. Cass & M. R. Wall	ASSE
An Illustrated Guide to Electrical Safety	OSHA	ASSE
Practical Electrical Safety	D. C. Winburn	ASSE
Behavioral Engineering through Safety Training: The B.E.S.T. Approach	J. P. Kohn	ASSE
Training in the Workplace	E. Heath & T. Ferry	ASSE
Applying Health & Safety Training Methods	J. P. Kohn & D. L. Timmons	ASSE
Safety by Objectives	D. Petersen	ASSE

Safety Training Methods	J. B. ReVelle	ASSE
Hazardous & Toxic Materials: Safe Handling and Disposal	H. H. Fawcett	ASSE
Safe Storage of Laboratory Chemicals	D. A. Pipitone	ASSE
Managing Safety & Health Programs	R. Boylston	ASSE
Techniques of Safety Management: A Systems Approach	D. Petersen	ASSE
Safety & Health Management Planning	T. Ferry	ASSE
Safety Management: A Human Approach	D. Petersen	ASSE
Handbook of Occupational Safety & Health	L. Slote	ASSE
Safe Behavior Reinforcement	D. Petersen	ASSE
Occupational Safety & Health Management	T. J. Anton	ASSE
Occupational Safety Management & Engineering	W. Hammer	ASSE
Risk Assessment & Risk Management for the Chemical Process Industry	Stone & Webster Engineering Co.	ASSE
Getting a Handle on Fire with Portable Fire Extinguishers	booklet	Factory Mutual Engineering Organization (FMEO)
Controlling the Power of Flammable Liquids	pamphlet	FMEO
The Dynamics of Dust Explosions	booklet	FMEO
Flammable Liquids-Risk, Regulations and Protection	booklet	FMEO
Construction Hazards	booklet	FMEO
Series of Guidelines and otherPublications		Scaffolding, Shoring & Forming Inst. (SSFI)

LP-Gas Safety Handbook		National Propane Gas Association (NPGA)
Safe Practices Around LP-Gas Installations	Bulletin	NPGA
First Aid for Propane Freeze	Bulletin	NPGA
Safety Bulletin on Use of Lifting Lugs on ASME Tanks	Bulletin	NPGA
Proper Clothing PromotesSafety	Bulletin	NPGA
LP-Gas Bulk Storage Safety Inspection Checklist	Bulletin	NPGA
LP-Gas Cargo Tank Truck Inspection Checklist	Bulletin	NPGA
Housekeeping for Safety's Sake	Bulletin	NPGA
Tank Car Unloading Safety Checklist & Return Procedures	Bulletin	NPGA
DOTs Hazard Identification for LP-Gas	Bulletin	NPGA
Reducing Static Electricity Fire Hazards	Bulletin	NPGA
How to Handle Small LP-Gas Fires with Portable Fire Extinguishers	Bulletin	NPGA
Guidelines for Conducting a Fire Safety Analysis	Bulletin	NPGA
Safety Considerations for Controlling Leaks and Fires in Propane-Powered Vehicles	Bulletin	NPGA
Safe Practices on Bulk Plants	Bulletin	NPGA
Safe Use of Hand Tools	Bulletin	NPGA
Safe Use of LP-Gas Cylinders in Outside Industrial Applications	Bulletin	NPGA
Safe Use of LP-Gas for Flame Cutting	Bulletin	NPGA

Safe Use of LP-Gas in Industrial Trucks	Bulletin	NPGA
How to Use LP-Gas Safely at Construction Sites	Bulletin	NPGA
Safe Use of LP-Gas with Portable Cylinders for Cutting & Brazing	Bulletin	NPGA
Safe Filling of Forklift Cylinders	Bulletin	NPGA
Risk Management Manual		NPGA
Accident Prevention Manual		National Safety Council (NSC)
Chemical Protective Clothing Performance Index		NSC
Laboratory Health & Safety Handbook: A Guide for the Preparation of a Chemical Hygiene Plan		NSC
NIOSH Pocket Guide to Chemical Hazards		NSC
Noise Control: A Guide for Employees & Employers		NSC
Supervisors' Safety Manual		NSC
The Right Start...Employee Safety Orientation	Booklet	NSC
Lockout/Tagout	Booklet	NSC
Electrical Safety at Work	Booklet	NSC
Hazardous Materials	Booklet	NSC
Good Housekeeping	Booklet	NSC
Lab Safety	Booklet	NSC
Four Sides of Danger	Booklet	NSC
Hand-Held Hazards	Booklet	NSC
Ladder Safety	Booklet	NSC
Using Tools Safely	Booklet	NSC

6.7.3 Audio-Visual Material

Title	*Organization*
Safety in the Chemical Process Industries	American Institute of Chemical Engineers (AIChE)
Recognition of Accident Potential	American Society of Safety Engineers (ASSE)
Introduction to System Safety	ASSE
Machine Tool Safety	ASSE
Why Eye Care	ASSE
Instructional Techniques for Safety Training	ASSE
An Approach to Industrial Explosion Protection	ASSE
Safer Operation of Screw Conveyors Three Basic Rules (Slides and Audio Tape)	Conveyor Equipment Manufacturers Association (CEMA)
Videotape Version of Above	CEMA
The A, B, C's and D's of Portable Fire Extinguishers	Factory Mutual Engineering Organization (FMEO)
Flammable Liquid Fire Safety	FMEO
Store Plastics	FMEO
Cutting & Welding Fire Safety	FMEO
Forming Safety Do's & Don'ts	Scaffolding, Shoring, & Forming Inst. (SSFI)
Shoring Safety Do's & Don'ts	SSFI
Suspended Powered Scaffolding Safety Do's and Don'ts	SSFI
Safety Guidelines for Frame Scaffolds	SSFI
Dispensing LP-Gas Safely	National Propane Gas Association (NPGA)
Handling LP-Gas Leaks & Fires	NPGA
LP-Gas Emergency Planning & Response	NPGA
Safe Use of LP-Gas Powered Lift Trucks	NPGA
Safety Management	National Safety Council (NSC)

Employment Safety Training	NSC
Safety Inspections	NSC
Industrial Hygiene	NSC
Personal Protective Equipment	NSC
Machine Safeguarding	NSC
Hand & Portable Power Tools	NSC
Materials Handling & Storage	NSC
Electrical Safety	NSC
Fire Safety	NSC
Job Safety Analysis	NSC
The Right Start...Employee Safety Orientation Slide Show	NSC
OSHA Lockout/Tagout	NSC
I&C Safety: Electrical Hazards	NSC
I&C Safety: High Temperature & High Pressure	NSC
Hazardous Materials Sampling	NSC
Process Sampling Safety	NSC
Low Voltage Safety	NSC
Electrical Equipment Hazards	NSC
Electrical Safety for Non-Electrical Workers	NSC
High Voltage Hazards	NSC
Protective Equipment in Electrical Maintenance	NSC
Safe Practices in Electrical Maintenance	NSC
Forklift Pre-Start Inspection & Safety Checks	NSC
Safe Forklift Operation	NSC
Decontamination	NSC
Introduction to Chemical Safety	NSC
Introduction to Hazardous Substances	NSC
Introduction to Reactive & Explosive Materials	NSC
Introduction to HAZWOPER	NSC

Introduction to HazCom	NSC
Material Safety Data Sheets (MSDS)	NSC
Hazardous Materials: Handle with Care	NSC
Industrial Hygiene	NSC
Safety Partner Good Housekeeping	NSC
Safety Partner Slipping, Tripping and Falling	NSC
Safety Partner Chemical Laboratory Safety	NSC
Research Laboratory Safety	NSC
Machine Safeguarding	NSC
Hand & Portable Power Tools	NSC
Safety Gear: Eye & Face Protection	NSC
Safety Gear: Foot Protection	NSC
Safety Gear: Hand & Arm Protection	NSC
Safety Gear: Head Protection	NSC
Safety Gear: Hearing Protection	NSC
Safety Gear: Respirators	NSC
Personal Protection	NSC
Fitting Respirator Equipment	NSC
Self-Contained Breathing Apparatus	NSC
Personal Protective Equipment	NSC
Benzene	NSC
Chlorine Safety	NSC
Sulfur-Based Gases	NSC

6.7.4 Training Kits

Title	*Organization*
Recognizing Ignition Sources: Fighting Fire Before It Starts	Factory Mutual Engineering Organization (FMEO)
Using Portable Fire Extinguishers: Five Basic Steps	FMEO

Preventing Cutting & Welding Fires	FMEO
Training Guidebook Package	National Propane Gas Association (NPGA)
WLGA Home Study Course	NPGA
Job Safety Analysis Instructor Manual	National Safety Council (NSC)
Job Safety Analysis Participant Manual	NSC
The Right Start...Employee Safety Orientation Instructors Manual	NSC
Confined Spaces: Training the Team	NSC

6.7.5 Computer Software

Title	*Organization*
ARCHIE	Federal Emergency Management Agency (FEMA)
REACT	Environmental Protection Agency (EPA)
FastRegs/OSHA	National Safety Council (NSC)
RegScan OSHA	NSC
RegScan RCRA	NSC
RegScan SARA	NSC
RegScan AIR	NSC
RegScan WATER	NSC
RegScan TSCA	NSC
Accusafe Software	NSC

REFERENCES

1. King, R., *Safety in the Process Industries*, Butterworth-Heinemann, London, 1990, p. 604, Fig. 21.1.
2. King, R., op. cit., p. 598.
3. National Safety Council, *Accident Prevention Manual for Business & Industry: Administration & Programs*, 10th ed., 1992, p. 370.
4. Ibid, p. 376.
5. King, R., op. cit., p. 615, Table 21.5.
6. National Safety Council, op. cit., p. 367.
7. Ibid, p. 368.

7

SPECIAL PROCEDURES AND PROGRAMS

Preceding chapters discussed the specifics of equipment safety and the proper execution of work within an operating unit. This chapter covers plantwide systems and more general programs whose purpose is to improve the overall safety performance of an organization. Such improvement can result from direct measures, such as process safety reviews or equipment inspections, or from more philosophical approaches that emphasize awareness and promote good attitudes toward safety. This chapter includes plant safety programs (7.1), industrial hygiene (7.2), voluntary industry programs such as the CMA's Responsible Care® initiative (7.3), and governmental regulatory programs (7.4).

7.1. PLANT SAFETY PROGRAMS AND AUXILIARY TOPICS

This section deals with some of the programs that can be used to improve a plant's safety performance. The first part briefly describes the most prominent techniques for process hazard analysis. A formal written program for analysis of certain facilities, with schedule for completion, is now required by OSHA (29CFR1910.119).

Other parts of this section discuss the management of change, emergency response plans, permit programs for hazardous operations, incident investigation, the use of safety manuals and incentive programs, and plant safety inspections.

7.1.1. Process Hazard Reviews

Safe operation of chemical processes requires a thorough analysis of safety hazards [1]. The resulting risks must then be evaluated against what is deemed acceptable. If the evaluation shows too high a risk, the hazard must be mitigated. The general subject of hazard analysis and the more prominent analytical techniques are covered thoroughly in the CCPS publication *Guidelines for Hazard Evaluation Procedures*. The most commonly used techniques are:

- Safety Survey
- Preliminary Hazard Analysis (PHA)
- "What-If" Analysis
- Hazard and Operability Studies (HazOp)
- Failure Modes and Effects Analysis (FMEA)
- Fault Tree Analysis (FTA)

Each of these methods is described briefly in this section. The CCPS volume mentioned above treats each of them in detail, covering the following aspects:

- purpose
- when to use
- type of results
- nature of results (qualitative or quantitative)
- data requirements
- staff requirements
- time and cost

A short summary of the characteristics of the more detailed methods is in Table 7-1. The general subject of process hazard management is also covered in API Recommended Practice 750.

7.1.1.1. Checklists and Safety Surveys
For simple operations, a chemical hazard survey could be simply an inventory of hazardous materials and a determination that they are stored safely, with workable plans to avoid fire or accident. With more complex operations, a more rigorous approach should be used. For example, the Dow Fire and Explosion Index (see Fig. 2-9) helps to rank hazards in relative terms by estimating potential damage [2].

Process checklists are also used for preliminary hazard identification. They are often developed in generic form for activities such as starting up a new process or modifying a current process. They should be followed by more detailed analyses.

7.1.1.2. Preliminary Hazard Analysis (PrHA)[1]
Preliminary Hazard Analysis is used in the early stages of design to identify hazards in broad terms. It considers

- chemical and physical properties of all materials
- interreactivity of the chemicals

---

1 The initials "PHA" are used in previous CCPS Guidelines for the Preliminary Hazard Analysis technique. In OSHA's Process Safety Management rule (29CFR1910.119), they refer to the general topic of Process Hazard Analysis.

- location of hazardous chemicals
- equipment
- operating conditions
- maintenance

Preliminary process flow diagrams and equipment arrangement drawings are reviewed when they are available. Serious hazards can be removed or mitigated before engineering work proceeds to an advanced stage.

With the PHA results, a decision diagram will identify the need for more detailed hazard analyses. Figure 7-1 is an example.

7.1.1.3. "What If" Analysis

In a "What If" analysis, a hazard review team poses hypothetical questions about potential process upsets or failures, then analyzes the consequences and proposes solutions. It is best applied when a major redesign of an existing plant is planned or when a new design is well documented and nearly firm.

The analysis team starts with a full set of process and instrument diagrams (P&IDs). The team considers possible abnormal events or combinations of events and reviews planned safeguards (e.g. interlocks, relief valves) for adequacy. The final report includes all "what-if" worksheets, a prioritized listing of identified hazards, and the modifications recommended to reduce the hazards to an acceptable level.

This method is popular because of its simplicity and suitability for group participation. However, it is not as rigorously structured as some of the techniques described below. To reduce the chance of overlooking a serious hazard, it is often combined with the use of checklists ("What-If"–Checklist method). Those using these methods should be aware of the tendency to regard checklists as complete and not to look for hazards not on the list.

7.1.1.4. Hazard and Operability (HazOp) Studies

HazOp studies are formal procedures used to identify and analyze hazards in a chemical operation. Like the "What-If" method, HazOp is inductive in its application, and it requires a full set of P&IDs and equipment specifications.

HazOp is well-structured and comprehensive. A review team rigorously applies a series of "guide words" to evaluate the process. Each process segment is examined for deviations from its design intention as directed by the guide words "no," "less," "more," "as well as," "reverse," "part of," and "other than." Each deviation is evaluated for resulting hazards. After consideration of existing safeguards, recommendations are made as appropriate to mitigate hazards to acceptable levels. Relevant details and recommendations are included in a final report.

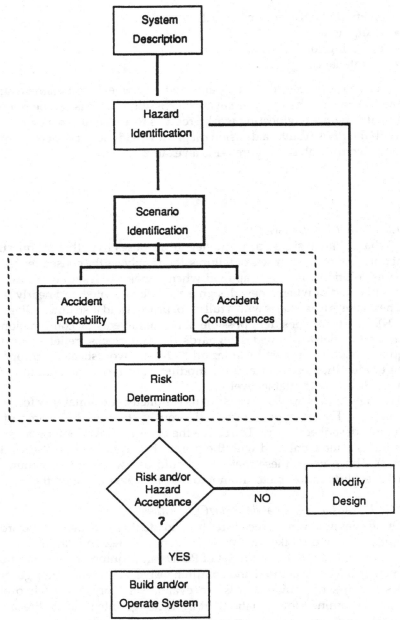

FIGURE 7-1. Preliminary Hazard Review Decision Diagram. (Adapted from *Guidelines for Hazard Evaluation Procedures,* AIChE, 1985, p. 1-9.)

This method is time-consuming because of its comprehensive nature. However, the structured approach minimizes the possibility of overlooking a process hazard.

7.1.1.5. Failure Modes and Effects Analysis (FMEA)

FMEA is used to evaluate the effects of the failure of a particular equipment item or component on other components and the entire system. Additional protective features and safety measures are then developed as required.

This method is equipment-oriented. It does not consider such problems as human error or raw material deviations, and it is not readily adaptable to complex systems. "What-If" or HazOp analysis is usually more successful in such cases. FMEA is best used for small segments of a process or repetitive analysis of similar equipment. Although two analysts are preferred in order to allow cross-checking, the method can be applied by an individual.

When the analysis includes evaluation of the criticality of the various failures, this method is referred to as Failure Modes, Effects, and Criticality Analysis (FMECA).

7.1.1.6. Fault Tree Analysis (FTA)

FTA is more complex than the above methods but is quite systematic and can be represented concisely. It is a deductive method that reveals the basic causes of a single known, undesired event. These causes can be equipment failures or human errors; the advantage of FTA is that it identifies problems that may be remote from their ultimate consequence and allows the analyst to focus preventive measures on them.

A fault tree shows a logical sequence of events that could result in a hypothetical or observed failure. A preliminary meeting first defines the major undesired events that are to be analyzed. At later sessions, each undesired event is placed at the top of a fault tree schematic. Figure 7-2 [3] shows the partial development of one side of a fault tree. The main objective with this method is to determine how failures or actions involving equipment, maintenance, or personnel could result in the undesired event at the top.

All team members should have a working knowledge of the process steps involved in the FTA and be able to participate in the construction of the fault tree.

FTA is most precise and thorough, but it is not a hazard identification method. It is used strictly for analysis of known major hazards for the purpose of mitigating those hazards.

7.1.1.7. Other Methods

The hazard analysis methods described above can be varied to suit circumstances, and many variations are beginning to appear. Other methods exist that are basically different and that are not specifically named by OSHA as satisfying the Process Safety Management rule (Section 7.4.2) Some of these are listed below.

TABLE 7-1
Hazard Analysis Methods

Description	Advantages	Disadvantages
"WHAT-IF" ANALYSIS		
• Method asks a series of questions. Each question evaluates a failure of a handling or process step. • Most often used to examine changes in a facility and for familiar or relatively uncomplicated processes..	• Relatively easy to use • Little training required. • Can be used for a wide range of hazards. • Compares process with past experience. • Effective learning tool for operations and more complicated hazard analyses.	• Sometimes used as an inexpensive shortcut when a more involved analysis would be more appropriate. • Depth of analysis is limited. • Effective only when the right questions are asked.
HAZARD AND OPERABILITY STUDY (HAZOP)		
• The most popular hazard analysis technique. • Requires a brainstorming multidisciplinary team of 5 to 10 people. The team identifies the consequences of deviations from the intended design of various process operations.	• Provides a methodical approach to the analysis of all deviations from the design basis. • Most applicable to chemical processes. • Easy to document. • Flexible. • Creative.	• Requires an accurate representation of the process system. Design drawings and specification must be available. • Assumes the process design is correct. • Must have a strong team leader to keep the group on track.
FAILURE MODES AND EFFECTS ANALYSIS (FMEA) also called Failure Modes, Effects, and Criticality Analysis (FMECA)		
• The method tabulates each system or piece of equipment together with its failure modes and the effects of each failure. The failures are then ranked by seriousness of resulting accidents.	• Provides a methodical approach to documenting failure modes and their consequences. • With proper training, easy to use and document. • Segments unusual processes for in-depth analysis.	• Requires an accurate representation of the process system. Design drawings and specifications must be available. • Assumes the process design is correct. • Focuses on instrumentation and equipment in "Go–No Go" situations.
FAULT TREE ANALYSIS (FTA)		
A formalized deductive technique that works backward from a defined accident to identify and graphically display combinations of equipment failure and operator error that can lead to the accident. Probabilities are assigned to each failure and error and an overall probability for the event is determined.	• Provides objective information for decision making. • Defines the specific routes that may cause an accident. • Assigns specific probabilities to each event that leads to the accident. • Analyzes human error.	• Special training and expertise required. • Not easily understandable to the casual reader. • The focus is on the accident rather than the process that leads to it.

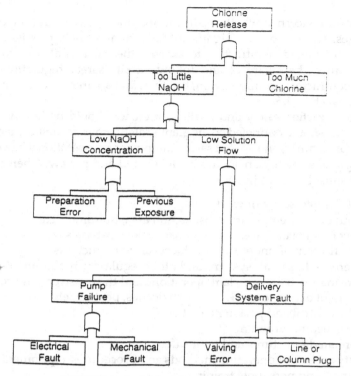

FIGURE 7-2. Partial Fault Tree—Chlorine Vent Scrubber.

Event Tree Analysis

This method is approximately the reverse of fault tree analysis. It begins with an "initiating event" and tracks its consequences. The initiating event may be a human error or a system failure.

Cause–Consequence Analysis

This method extends other analyses; it is a blend of fault tree and event tree analysis. Basic events can be linked not only to the resulting accident but also to its deleterious effects.

Human Error Analysis

Human Error Analysis encompasses a number of different techniques. All should relate tasks to the skills required to perform those tasks.

7.1.2. Facility and Operational Changes

Plants are often upgraded, modified, or reconstructed in some manner. Examples range from minor equipment modifications to large scale demolition and construction projects.

One major concern is whether current operations must continue during such changes. This can present significant hazards and safety problems.

The use of outside contractors to execute the work is also of concern. Contractors must be aware of and follow all site safety regulations. The subjects of contractor communication, training, and awareness of hazards are discussed in Section 4.6.

A preconstruction safety and health conference should be held with the contractor, who should be briefed on the plant's safety program, first-aid facilities, emergency procedures, and special safety equipment for operational hazards.

An implementation plan should be developed and reviewed before construction begins. It should include

- well-defined scope of work
- layout of site, temporary construction area, buildings, etc.
- delivery and staging areas for construction materials
- identification of underground objects and obstructions
- adherence to permit systems and other regulatory requirements
- coordination between plant operations and the construction groups
- control of access to work areas (barricades, posting, etc.)
- traffic control and parking facilities
- housekeeping program
- job-related structures and their safety
- safety aspects of operating methods and procedures, equipment maintenance, and personnel training
- fire protection
- ventilation
- protective equipment and apparel
- medical care
- definition of responsibilities

Plant employees and equipment must be protected from construction hazards, including excavations, falling objects, welding operations, dust, dirt, and the shortcomings of temporary electrical service. Construction areas should be isolated by signs and barricades, fences, guardrails, or fully sealed bulkheads.

Only fully inspected and properly maintained equipment should be allowed on site. This applies to power shovels, concrete mixers, supply trucks, cranes, materials handling devices, power tools, and other construction equipment. Great care must be taken to avoid collisions of mobile equipment with production equipment, pipelines, power lines, staged materials, or other objects or hazards. Specific measures include the use of permit programs, warning signs, personnel training, site reviews, and careful supervision.

Other safety-related issues include:

- For equipment
  - —guards for all moving parts
  - —electrical grounding
  - —engine exhausts directed outdoors
  - —platforms and accessways
  - —devices to prevent unauthorized activation
  - —no accumulation of debris
  - —refueling and handling of flammable liquids
  - —repair procedures and lockouts
- For construction
  - —full safeguards for steel erection
  - —lateral bracing for incomplete structures
  - —temporary flooring
  - —torches and furnaces
  - —temporary and portable heaters
  - —fire protection
  - —demolition
  - —hoists, formwork, and falsework
  - —floor and wall openings
- For excavation
  - —safeguards for underground utilities
  - —barricades
  - —trench bracing
- For working surfaces:
  - —use of ladders
  - —scaffolds and equipment
  - —platforms and other workways

7.1.3. Emergency Response

Advance planning for potential emergencies will help to avoid personal injury or damage to property. A comprehensive emergency management plan will allow quick and effective response and so reduce the consequences of any incident. Key actions include:

- Identify potential emergencies.
- Assess possible damage.
- Estimate available response time.
- Prepare plan.
- Publish and disseminate plan.
- Practice the plan.

7.1.3.1. Potential Emergencies
Emergencies can arise from many causes, for example:

Internal	External
fire, explosion	storm, hurricane, tornado, flood, earthquake, lightning, extreme temperature
hazardous or radioactive material	off-site spills, fires, explosions, etc.
work accident	accidents in nearby plants
shutdown	crime, civil disturbance
crime, bomb scare, sabotage	other: area fire, dam break, transportation stoppage or cutoff, utility outage or accident, etc.
toxic release	
medical emergency	
strike	
rumor	

Those with reasonable probabilities of occurring must be identified and specifically planned for. Government and private sources can help to identify likely external events.

7.1.3.2. Assessment of Damage
For each situation, the risk of damage to personnel, property, and the environment must be assessed. This assessment enables the planners to set priorities by considering:

- primary danger
- damage or hazardous circumstances resulting from cascading effects (e.g., a hazardous release from a tank caused by a nearby fire)
- time of day and workshift patterns
- nonstaffed or minimally staffed periods

7.1.3.3. Available Response Time
Each hypothetical emergency will demand a response within a certain time. Some examples are:

- immediate action—fire and explosion
- action within minutes—bomb threat
- action within hours—flood
- action within days—hurricane

7.1.3.4. Emergency Response Management Plan
A plan can be narrowly focused to deal only with responses during emergencies or can include broader actions before and after the events. Such plans can include:

- Emergency command structure, personnel requirements, communications, and resources needed (e.g., alarm system, medical treatment materials, auxiliary power supply).
- Emergency response approach, including organization and training of response teams. Small, well-trained teams are the first line of defense against disasters and can be used for evacuation, fire fighting, rescue, spill response, first aid, and other tasks.
- Specific procedures for emergency shutdown and evacuation.
- Evacuation routes and procedures.
- Follow-up efforts such as cleanup, salvage, overhaul, decontamination, and restoration of operations.
- Contingency plans when help from outside agencies is not available.
- Mutual aid arrangements with neighbors.

7.1.3.5. Dissemination and Practice of Plan

An emergency response plan will be effective only if it is clearly understood by all participants. This will happen if the plan is well-conceived, clearly written, made available to the interested parties, and then practiced. The benefit of drills can not be overemphasized. The written plan should address:

- company policy, purposes, authority, principal control measures, and emergency organization chart showing positions and functions
- description of potential emergencies with risk factors
- maps of plant and offices showing equipment, medical and first aid centers, fire control apparatus, shelters, command center, evacuation routes, and assembly areas
- central communications center, with information regarding cooperating agencies and emergency contacts
- plant warning system
- visitor and customer handling
- emergency equipment and resource listings and locations
- specific procedures for response to defined events, including shutdown procedures for operation and support functions
- training plans and schedules

While not directly safety-related, responsibility for communication with news media should be set. The relevant parts of the plan should be made available to the public, in accord with CAER or the Responsible Care® initiative (Section 7.3).

7.1.4. Permits

Certain permits and licenses are required by law before a plant is permitted to operate. These might indicate, for example, that fire prevention codes [4] or local regulations have been satisfied. This section is concerned not with such permits but with those issued on site to ensure that certain work is performed safely. Examples of work that should be performed only with a specific permit are

- entry into a vessel or a confined space
- work on rotating machinery
- breaking into lines or removing components
- opening of vessels or other equipment
- hot work
- work with especially hazardous materials
- electrical work
- bypassing of interlocks or other safety devices and instruments
- vehicle operation
- disposal of hazardous material
- excavation
- after-hours work
- sprinkler valve closing

Many tasks will fall into more than one of these categories. A plant may also have a work-order system or one that requires specific authorization for all maintenance work. This system should be written in a plant operating or maintenance procedures manual. It should be prominent in safety training programs.

While each hazardous work situation requires certain special considerations, there are also some overall guidelines that should be captured in a permit document. These guidelines include the need to

- define the work to be done, the location in which it is to be performed, and the operations that will be affected
- establish an approximate schedule and limit the time of validity of the permit
- establish responsibility for safety of work, including any transfer of responsibility during a shift change
- define procedures during new emergencies (e.g., secure work area and suspend permit)
- secure necessary approvals
- suspend other work in the vicinity of the permit activity
- inspect the area before starting work
- make all necessary tests (e.g., for oxygen, toxic materials, flammable vapors, tight joints)
- remove, correct, or clearly mark all hazardous materials and conditions
- prepare the site
- notify all personnel in and around the area of the work
- restrict access to the work area
- explain precautions and emergency measures to be taken
- provide all necessary safety, emergency, and personal protective equipment (e.g., fire extinguishers, rescue lines, gas masks)

- be sure that assigned personnel can use such equipment properly
- notify appropriate departments and authorities, e.g., plant or local fire department
- assign fire watch personnel during hazardous work or interruption of normal protection
- define workforce requirements
  —qualifications
  —manhours
  —schedule
  —duration of work
- assemble tools, work materials, and equipment of material to be installed before beginning work
- certify and label all material as suitable for the work
- separate combustible work materials from all ignition sources
- arrange safe disposal of hazardous materials and all trash generated by the work
- establish the electrical classification of the area
- inspect the area after work is completed, making sure it is in proper condition after cleaning and removal of scrap and tools
- test safety systems before resuming normal operation.

Specific types of permits are discussed in this section, and samples of various specialized permits have been published [5].

Hazardous Operation Permits

Hot Work (Welding, Cutting, Open Flame)
A hot-work permit is usually required for operations that use flames or produce sparks. The permit is intended to prevent fire or injury to the person doing the work. It should require the responsible party to:

- inspect the work area for hazards
- mark off the area to restrict access
- assign a person to watch for fire, if sufficiently hazardous, and maintain the fire watch for thirty minutes after the spark- or flame-producing equipment is shut down
- locate fire extinguishing apparatus at the site, manned by a standby employee
- coordinate efforts of all departments concerned with fire protection
- isolate combustible materials from all ignition sources
- verify by test that the work area is free of flammable vapor; retest as appropriate
- control the use of all spark- or flame-producing equipment
- actively manage any hot work that is performed by outside contractors

Electrical Hot Work

Electrical maintenance work should not normally be performed on live circuits. A work permit should be issued whenever electrical maintenance work is done in a hot (operating process) location. The permit should require the following points to be addressed:

- possible use of alternative or temporary circuit to allow process continuity
- effects of temporary work on normal operation
- effects of shutdown of circuits on vital functions such as pressurization equipment, alarms, etc.
- conformance with NEMA classification of enclosures, area classification, short-circuit capability, voltage and current capacity, etc.
- grounding connections available for portable tools
- type and location of warning signs

Confined Space Entry

Since the characteristics that define a permit-required space may come and go with changes in equipment configuration or in the process itself, classification can change. When a previously exempt space acquires one of the defining characteristics listed in Section 4.5.1, the employer must reclassify it as a permit-required space. Conversely, if a change removes all criteria, the space can be downrated. To do this, the employer must satisfy paragraph (c)(7) of the OSHA rule (29CFR1910.146). In a multipurpose unit, frequent changes are, in principle, allowed. Constant change, however, is confusing and not sound policy.

When an employer's survey of his plant shows that there are spaces that come under OSHA's rule, the employees must be so informed and the spaces posted. The owner may choose not to allow entry into these spaces, in which case he must take effective measures to prevent employees from entering. Should the employer decide that entry is necessary or desirable, he must develop a written permit space entry program.

An employer has similar obligations toward outside contractors whose services are used in the plant. The contractors must be

- informed that such spaces exist
- told of any hazards present
- told what precautions must be taken and what procedures must be followed for entry
- debriefed after any entry operation regarding the hazards encountered during the work and the effectiveness of the program

The owner must coordinate activities when both inside and outside personnel are involved. The contractor has the obligation to

- obtain relevant information on hazards and entry requirements from the owner
- inform the owner of the permit program that the contractor intends to use
- coordinate operations with the owner

For any permit-required space, the owner is obliged to

- implement all measures necessary to prevent unauthorized entry
- identify and evaluate hazards before permitting entry
- develop means and procedures for safe entry operations; as a minimum:
  —specifying acceptable conditions for entry
  —isolating the space
  —purging, flushing, inerting, or ventilating the space to control hazards
  —providing pedestrian and vehicular barriers as necessary to protect entrants from external hazards
  —verifying that conditions within the space are acceptable throughout the operation
- provide, maintain, and ensure the safe use of the following:
  —testing and monitoring equipment
  —ventilating equipment
  —communication equipment
  —personal protection equipment
  —lighting (for safe work and quick exit)
  —barriers and shields
  —equipment for safe ingress and egress (e.g., ladders)
  —rescue and emergency equipment, except as provided by outside rescue services
  —any other equipment necessary for the specific space and tasks being performed
- evaluate permit space conditions
  —check for acceptable conditions before entry
  —monitor space during work
  —test atmosphere, in sequence, for oxygen, combustibles, and toxic materials
- provide attendant(s) outside permit space for duration of operation; if one attendant is responsible for more than one space, adopt means and procedures to allow response to emergency in one space without default on the others
- designate those with active roles in entry procedure (e.g., authorized entrants, attendants, supervisors, testers), define their duties, and provide all necessary training
- develop and implement rescue procedures
- coordinate operations when employees of more than one employer work simultaneously within a permit space so that one group does not endanger another
- develop and implement procedures for concluding entry and closing off the permit space when the authorized work is complete

When the only hazard is an actual or potential hazardous atmosphere and it has been demonstrated that continuous forced-air ventilation is sufficient to keep the enclosed space safe for entry, there is an alternative procedure that is less strict. This can be found in paragraph (c)(5)(ii) of the OSHA rule.

An entry permit by law must identify

- the space to be entered
- reason for entry
- date of entry and the authorized duration
- the authorized entrants[2]
- name(s) of the attendant(s)
- name of the supervisor
- signature or initials of the person who originally authorized entry
- the hazards that will be encountered
- methods used to isolate space and control hazards
- conditions that must be achieved before entry
- results of tests, accompanied by names or initials of the testers and an indication of when the tests were performed
- available rescue and emergency services
- procedure for summoning the above
- procedures for communicating with attendant(s)
- equipment required (PPE, testing equipment, alarm systems, communications equipment, rescue gear)
- other necessary information specific to task
- other permits already issued for work in the space.

The permit form often is supplemented by a pre-entry checklist. Published examples are in 58 FR 4560-2.

Specialized Permits
A plant may choose to regulate certain types of work by the use of other permits. Some of the activities for which they may be appropriate are listed below:

Hazardous Operation
- *Line breaking (opening)* Opening of pipelines, pumps, or other attached apparatus.
- *Waste disposal* Permit or license may be required for disposal of any waste; only hazardous wastes are of interest here.
- *Excavation* Contaminated soil; unstable soil conditions; underground obstructions, pipelines, or power lines.
- *After-hours work* Unusual operations outside of normal working hours; unattended operation. Responsibility on- and off-site should be established, and emergency contact names and telephone numbers should be available to security personnel.

---

2 A tracking system must be used to monitor all entrants in and out of the confined space.

- *Vehicle Operation* Issued to operator after special training which includes plant rules, operating procedures, and a realistic test; limited duration before requalification.
- *High work* Work on scaffolds, platforms, pipe racks, or elevated equipment.

Hazardous Area Entry
- *Explosives building* Building containing explosives, propellants, or pyrotechnic material.
- *Corrosives area* Corrosive material manufacturing or processing area.
- *Flammables area* Area in which large amounts of flammables are processed or stored.

Special Hazards
- *Toxic materials* Handling procedures, material control, and personnel qualification.
- *Matches and lighters*
- *Vehicle operation* Specialized vehicles such as fork lift trucks, payloaders, graders, bulldozers, transporters, and hoists.
- *Sprinkler valve closing* Shutdown of fire protection system by closing fire main or sprinkler system (note: this may be covered by local fire or insurance company regulations).

Many of these activities will also require the sorts of permits already discussed. In such a case, a single document may suffice.

7.1.5. Incident Investigating and Reporting

In an ongoing safety program, incidents and accidents are investigated and analyzed in order to reduce the likelihood of recurrence. The detailed information gained from these investigations is used to develop corrective action.

An assigned group should investigate both personal injury and property damage incidents. The seriousness or potential of the incident will dictate the extent of the investigation. Such an investigation identifies and locates principal hazards and outdated or inefficient processes and procedures. The analysis should produce recommendations for changes to work methods, equipment, processes, or training programs. This effort can be assisted by the detailed checklist "Guide for Identifying Causal Factors and Corrective Actions" which is available from the NSC.

Accident reports and recordkeeping are required by OSHA and other local and state agencies. Companies must have written policies and generally have standard forms for reporting, which include employer data, employee characteristics, narrative description of incident, associated equipment, task being performed at the time, time setting, protective clothing and training, and description of the injury or damage. ANSI Standard Z16.2 is widely used. The policy should also require a report on follow-up action.

The analysis of an accident often includes an estimate of the cost of that accident. This helps in justifying the cost of accident prevention activities. Accidents are generally divided into four classes:

- fatalities
- lost-time injuries
- recordable injuries
- near-misses

An accident will have both insured and uninsured costs. Insured costs include worker's compensation insurance premiums and medical expenses. Uninsured costs include:

- damage to material or equipment (including reorganization of material or resetting of equipment)
- lost wages for injured workers beyond those covered by insurance
- lost time by other workers responding to accident
- overtime wages (making up lost production, extra supervision, heat, light, cleaning, etc.)
- supervisory wages for accident-related time (time spent away from normal tasks)
- continuing wages for injured worker's decreased output (reduced capability while returning to full health)
- additional wages for replacement worker (training time, lower efficiency, extra supervisory time)
- medical costs (dispensary usage)
- administrative costs (accident investigation time, paperwork generation, claim administration, including both management and clerical time)
- other costs (equipment rental, lost orders or canceled contracts from nondelivery, liability claims, etc.)

7.1.6. Safety Manuals

Each plant needs a manual that documents rules and guidelines for safe operation. The manual should also contain relevant health and environmental information.

Procedures should clarify the hazards in operations and discuss the consequences if safe limits are exceeded. The procedures must be kept up-to-date, and personnel should be trained accordingly.

Information in a safety manual usually includes:

- accident prevention policy
- accident reporting and investigation procedures (7.1.5)

- first aid
- medical assistance
- safety equipment availability, with locations
- housekeeping
- waste disposal
- personal protection (4.7.1)
- smoking and spark or flame generation
- emergency response plan (7.1.3)
- permit procedures (7.1.4)
- communication
- fire prevention, detection, and reporting
- welding, maintenance, and inspection
- other occupational safety and health information

A separate manual may describe the safety aspects of operating procedures:

- organizational responsibility and authority in each operating area
- unloading, loading, and material handling procedures
- startup and shutdown procedures
- normal, temporary, and emergency operating conditions and safety limits
- consequences of deviations, with steps to avoid or correct them
- process safety systems and their functions
- use of equipment and tools, including material handling devices
- other operating rules and procedures

7.1.7. Safety Incentives

Workplace safety is enhanced when:

- Management is committed to and actively supports a substantive safety program.
- The workplace has been designed for safety.
- Procedures have been made as safe as possible.
- Supervisors have trained their crews thoroughly and continue to monitor procedures.
- People want to work safely.

The first four points are the subject of most of this volume. The last is essential if the others are to be effective. Employees must think for themselves in order to act safely in hazardous situations. A good "safety atmosphere" will encourage the use of common sense, self-discipline, and imagination. A good safety program that stimulates and maintains employee interest in safety can involve

- education (publications, videos, films)
- publicity (posters, memos, displays)
- training

- accident analysis
- communication methods (meetings, questionnaires, committees, quality/safety circles)
- incentives

Incentives might include individual group recognition, publicity, money, and merchandise awards. There are many creative ways to motivate employees. Safety newsletters and publications describe many of these, and safety professionals and employees themselves are rich sources of ideas.

7.1.8. Inspections and Reporting of Unsafe Conditions

Inspections help to uncover existing and potential unsafe conditions. Inspections should focus on fact-finding, not fault-finding, so that hazards can be co-operatively identified and corrected before an accident occurs. They can also help to improve operations and so produce higher efficiency, effectiveness, and profitability.

Both formal and informal inspection programs are important. These programs, the types of items to be inspected, and methods for inspection are described below.

7.1.8.1 Scheduled Inspections

Planned inspections follow established procedures and use checklists for routine items. They can be periodic, intermittent, and general.

Periodic inspections are scheduled at regular intervals (e.g. weekly, monthly, annually), generally more frequently when the hazard potential is higher. They include:

- plant safety inspections for fire extinguishers, safety guards, and other safety devices
- inspections by maintenance personnel for mechanical functions, lubrication, grounding, etc.
- outside inspections by certified or licensed inspectors of boilers, elevators, pressure vessels, cranes, power presses, sprinkler systems, etc.
- outside inspections by government or regulatory personnel to establish compliance with regulations

Intermittent inspections are made at irregular intervals, usually in response to specific needs or occurrences. Examples are:

- accident investigations
- higher-than-normal accident or illness rates in a specific department
- local construction or renovation
- installation of new equipment
- process changes
- suspected environmental or health hazards.

General inspections cover areas infrequently inspected, such as parking lots, sidewalks, fencing, and grounds. Overhead inspections can locate problems with skylights, windows, roofs, or cranes. They are also required before reopening a plant after a long shutdown.

7.1.8.2 Routine Inspections

These include continual ongoing inspections by employees as part of their normal duties. Such inspections involve noting hazards or unsafe procedures and correcting them or reporting them for corrective action.

Supervisors and foremen make sure that tools, machines, and equipment are properly maintained and safe to use, and that safe procedures are followed in their areas. Workers observe equipment and its functioning, as well as other general conditions within the plant.

The most difficult hazards to identify are those which result from poor housekeeping and unsafe practices which are caused by gradual deterioration and accumulation over time. Personnel from another area may have to help with this type of inspection.

7.1.8.3 Subjects for Inspection

Any situation or equipment that may become unsafe or cause accidents should be identified and inspected. Examples are:

- *Environmental Factors:* Illumination, dust, fumes, gases, mists, vapors, noise, vibration, heat, radiation.
- *Hazardous Materials:* Explosive, flammable, acidic, caustic, toxic, radioactive.
- *Production and Related Equipment:* Pumps, agitators, vessels, piping, machine tools, specialized equipment.
- *Electrical Equipment:* Switches, fuses, breakers, outlets, cables, power cords, grounds, connectors, connections.
- *Power Equipment:* Engines, motors, turbines.
- *Hand Tools:* Power tools, wrenches, screwdrivers, hammers.
- *Personal Protective Equipment:* Hard hats, safety glasses, safety shoes, respirators, hearing protection, gloves.
- *Personal Service and First Aid Facilities:* Drinking fountains, wash basins, safety showers, eyewash fountains, first aid supplies, stretchers.
- *Elevators and Lifts:* Controls, wire ropes, safety devices.
- *Material Handling Equipment:* Cranes, dollies, conveyors, hoists, forklifts, chains, ropes, slings.
- *Transportation Equipment:* Automobiles, railroad cars, trucks, front-end loaders, motorized carts and buggies.
- *Fire Protection and Emergency Response Equipment:* Alarms, water tanks, sprinklers, standpipes, hydrants, hoses, self-contained breathing apparatus, automatic valves, horns, phones, radios.

- *Warning and Signaling Devices:* Sirens, crossing and blinker lights, klaxons, warning signs, exit signs.
- *Containers:* Scrap bins, disposable receptacles, carboys, barrels, drums, gas cylinders, solvent cans.
- *Storage Facilities:* Bins, racks, lockers, cabinets, shelves, tanks, closets.
- *Buildings and Structures:* Floors, roofs, walls, fencing.
- *Structural Openings:* Windows, doors, stairways, sumps, shafts, pits, floor openings.
- *Working Surfaces:* Ladders, scaffolds, catwalks, platforms.
- *Walkways and Roadways:* Ramps, docks, sidewalks, walkways, aisles, vehicle ways, escape routes.
- *Employee Behavior:* Operation of vehicles, use of machinery and tools, use and maintenance of PPE, sanitation, hygiene, smoking, housekeeping, work habits, observance of precautions and use of safety devices, materials handling, lifting.

Those items most likely to become serious hazards deserve particular attention. Problems can be aggravated by stress, wear, impact, vibration, heat, corrosion, chemical reaction, or misuse.

7.1.8.4 Inspection Methods
Checklists are frequently used to guide inspections. When comprehensive and well thought out, their use avoids potentially serious omissions. Many samples are available from OSHA and other safety organizations [6]. Checklists can be used directly or modified to meet specific needs. They are also valuable in follow-up work, to verify that corrective actions have been taken and that the hazardous conditions have been eliminated or minimized.

The optimum frequency of inspection depends on four factors:

- potential severity
- potential for injury
- speed with which an unsafe condition can develop
- past history of failures

These factors, along with company and government regulations, will set the schedules for inspection. Qualified personnel should carry out the inspections. They must have the appropriate tools and protective equipment and be given time to prepare and plan their work.

Inspectors should locate hazards, identify them in detail, and classify or rank them by seriousness. Immediate action, including equipment shutdown, is appropriate when there is imminent danger. Comprehensive inspection reports allow management to understand and evaluate each situation, so that corrective action decisions can be made promptly.

Some guidelines for inspection reports are:

- Describe hazards and equipment conditions explicitly.

- Address the cause whenever possible, not the result.
- Make immediate corrections where possible in order to avoid future accidents.
- Report conditions beyond immediate correction and suggest solutions.
- When solutions will take some time, recommend interim measures to mitigate hazards.
- Maintain records to identify trends; measure progress.

7.2. INDUSTRIAL HYGIENE

Environmental factors or stresses in the workplace can cause sickness or significant discomfort among workers or citizens of the community. Proper industrial hygiene will alleviate these effects. It includes the identification of health hazards, the evaluation of their magnitude, and the development of corrective measures to control them. The goals of an industrial hygiene program are to:

- protect workers from health hazards in the workplace
- ensure that people are physically, mentally, and emotionally capable of performing their jobs efficiently and safely
- assure proper medical care and rehabilitation of individuals who have been injured on the job
- encourage employee health maintenance

The purposes of the OSHAct are to assure safe and healthful working conditions and so to preserve human resources. Details of OSHA requirements are covered elsewhere in this book. However, nearly every employer is required to implement some portion of an industrial hygiene or occupational health program. The primary benefits of such a program are:

- improved employee health
- reduced compensation costs
- increased productivity and efficiency
- improved product design and packaging
- improved process design
- reduced labor turnover
- improved labor relations

An industrial hygiene program needs participation by several organizational elements. These elements and their primary responsibilities are listed in Table 7.2.

Results should be measured in order to determine how well an industrial hygiene program is working. Table 7-3 lists various phases of a program and some measures of its performance [7].

TABLE 7-2
Elements of Industrial Hygiene Program: Responsibilities of Various Groups

MEDICAL
—Help develop effective measures to avoid exposure to hazardous materials or situations
—Periodically examine employees exposed to hazardous materials
—If indicated, restrict employees from further exposure to hazardous materials

INDUSTRIAL HYGIENE
—Identify potential hazards in processes and operations (current, new, and planned); notify
  responsible organizations
—Review all work practices for conformance to standards
—Establish hygienic standards and periodically test for compliance
—Specify design/quality of personal protective equipment; define standards for use
—Recommend controls to minimize exposure to environmental hazards
—Assist in safe work practices training program

ENGINEERING
—Plan operations to prevent exposure to hazardous materials or situations
—Notify hygiene, medical, safety organizations of planned process or operation changes
—Request survey by industrial hygiene of new installations before starting operations

SAFETY
—Conduct effective safety program; coordinate related educational, engineering, supervisory,
  enforcement activities
—Conduct safety training for employees working with hazardous materials
—Assist in safe work practices training program
—Conduct safety surveys to assure compliance with safe practices and procedures
—Recommend changes in safety practices and procedures as technology and information change

PURCHASING
—Ensure that equipment and materials purchased are approved by medical, industrial hygiene,
  safety, and other appropriate departments

SUPERVISORS
—Maintain safe work environment for employees
—Ensure that employees are examined and approved by the medical organization before working
—Conduct safe work practices training program; assist in safety training for employees working
  with hazardous materials
—Maintain good housekeeping practices
—Promptly inform engineering, industrial hygiene, and safety organizations of potentially
  hazardous situations
—Promptly inform medical organization of accident or exposure to hazardous materials; send
  employee for medical examination
—Furnish proper protective clothing and training to those working with hazardous materials or in
  dangerous situations
—Observe all work restrictions imposed by the medical organization; administer appropriate
  disciplinary action when health and safety practices and procedures are violated

EMPLOYEES
—Promptly inform supervisor of any hazardous situations, materials, processes, or operations, or
  any accidents or exposure to harmful materials
—Observe all safety practices and procedures; use personal protective clothing when necessary
—Practice good housekeeping and personal hygiene

TABLE 7-3
Industrial Hygiene Performance Measures

CONTROL OF OCCUPATIONAL HEALTH HAZARDS

A. *Program Measurement*
1. Identification of health hazards
2. Hazard evaluation
3. Control measures in use
4. Periodic monitoring

B. *Measurement of Results*
1. Employee health complaints.
2. Labor turnover.
3. Employee morale.
4. Dispensary visits.
5. Occupational disease cases.
6. Workers' O.D. compensation costs.
7. Sick absenteeism.

ENVIRONMENTAL CONDITIONS AFFECTING COMFORT

A. *Program Measurement*
1. Evaluation of factors affecting comfort
2. Control measures in use
3. Periodic monitoring

B. *Measurement of Results*
1. Employee productivity.
2. Employee complaints on comfort conditions.
3. Employee morale.

ENVIRONMENTAL CONDITIONS AFFECTING JOB EFFICIENCY

A. *Program Measurement*
1. Identification of factors affecting physiological response
2. Evaluation of factors
3. Control measures in use
4. Periodic monitoring

B. *Measurement of Results*
1. Employee productivity.
2. Incidence of employee fatigue.
3. Number of product defects.
4. Accident frequency rate.

HEALTH HAZARDS OF THE FIRM'S PRODUCTS

A. *Program Measurement*
1. Eliminate hazards in products
2. Education of users
3. Toxicological information
4. Proper labeling

B. *Measurement of Results*
1. Review of product liability.

AIR POLLUTION CONTROL

A. *Program Measurement*
1. Compliance with regulations
2. Adequacy of control facilities
3. Periodic monitoring

B. *Measurement of Results*
1. Neighborhood complaints.
2. Property damage claims.
3. Frequence and size of releases.

CONSULTATION ON WORKERS' COMPENSATION CLAIMS

1. Analysis of background
2. Collection of facts on exposure

Support to Professional Societies and Associations
1. Assist in developing data sheets, standards, guides, etc.

Educational Activities.
In-Plant, community.
Assistance to legislative bodies developing codes, etc.
Communication with and Assistance to Other Disciplines.

7.3. VOLUNTARY PROGRAMS

7.3.1. Responsible Care®

The Responsible Care® program promotes safety in the manufacturing, transportation, use, and disposal of chemicals. It also emphasizes health and environmental considerations and compliance with governmental regulations and standards. Responsible Care includes a commitment to involve the public directly in deliberations and to work actively with suppliers, shippers, packagers, and end-users to enhance safety in all stages of a chemical's life.

While Responsible Care as conceived is a voluntary program, the CMA, whose members control more than 90% of the basic chemical capacity in the United States and Canada, has made participation in this initiative a requisite for membership. Each CMA company, and each of its chemical businesses, must subscribe to the principles of Responsible Care and is expected to participate in the development of programs.

The official guiding principles require a subscriber to:

- recognize and respond to community concerns about chemicals and chemical plant operations
- develop and produce chemicals that can be manufactured, transported, used, and disposed of safely
- make health, safety, and environmental considerations a priority in planning for all existing and new products and processes
- report promptly to officials, employees, customers, and the public information on chemical-related health or environmental hazards and recommend proactive measures
- counsel customers on the safe use, transportation, and disposal of chemical products
- operate plants and facilities in a manner that protects the environment and the health and safety of employees and the public
- extend knowledge by conducting or supporting research on the health, safety, and environmental effects of products, processes, and waste materials
- work with others to resolve problems created by past handling and disposal of hazardous substances
- participate with government and others in creating responsible laws, regulations, and standards to safeguard the community, workplace, and environment
- share experiences and offer assistance to others who produce, handle, use, transport, or dispose of chemicals

Involvement of the public is through the Public Advisory Panel (PAP), a diverse group of about fifteen individuals from both private and public sectors. An independent facilitator appointed by CMA selects the PAP members. The panel offers comments and advice, identifies issues that require industry response, and provides early definition of public concerns that affect the chemical industry.

In conjunction with the Public Advisory Panel, the CMA has developed a series of Codes of Management Practice for the Responsible Care partners. The first Codes address community awareness and emergency response, distribution, waste and release reduction and management, safe plant operation, and product stewardship. They set forth expected management practices. Quantitative goals are left to the member companies. The Codes thus complement existing programs and challenge the subscribing companies to achieve constant improvement.

The Process Safety Code was adopted in 1990 and is designed to reduce the risk of fires, explosions, and accidental chemical releases. It requires plants to:

- have an ongoing process safety program that includes measurement of safety performance, audits, and implementation of corrective actions
- conduct thorough safety reviews on all new and modified facilities during design and before startup
- conduct and document maintenance and inspection programs to ensure soundness of facilities
- develop and put in place sufficient layers of protection to prevent escalation from a single failure to a catastrophe
- train all employees to reach and maintain proficiency in safe work practices and the skills and knowledge necessary to perform their jobs
- share relevant safety knowledge and lessons learned from incidents with industry, government, and the community to help others to avoid similar incidents.

The Process Safety Code also requires that each safety program cover contractors' employees. Accidents have resulted from their lack of familiarity with plant hazards and from lack of communication during times of change or emergency. Under the Process Safety Code, a contractor must have a safety program consistent with the Code or must have his employees included in the host company's program.

The Process Safety Code, like all Responsible Care Codes, encourages community involvement and education of the public by the operating chemical company.

The first code adopted, and one of the most familiar to the public as well as to industry, was adapted from CMA's Community Awareness and Emergency Response (CAER) program. Its primary aim is to reduce the risk of injury to employees and neighbors in case of a plant accident. CAER required subscribing companies to:

- develop emergency plans
- communicate them to providers of emergency services and to the public
- test them annually

It also encouraged them to exchange information with each other and with local residents. CAER has been a successful concept. Many of its elements have become law (e.g., the "right-to-know" parts of SARA Title III), and it has been one of the foundations of the Responsible Care program.

CMA can provide detailed information on Responsible Care. *Chemical Week* [8] has published a worldwide review and is continuing periodic updates [9].

7.3.2. Service and Trade Associations

Some associations available to assist industry in the area of plant safety are listed below. These include insurers, technical societies, and standards organizations.

American Gas Association (AGA)
1515 Wilson Boulevard
Arlington, VA 22209

American Iron and Steel Institute (AISI)
1133 15th Street, NW
Washington, DC 20005

American National Standards Institute (ANSI)
11 West 42nd Street
New York, NY 10036

American Petroleum Institute (API)
1220 L Street, NW
Washington, DC 20005

American Society for Testing and Materials (ASTM)
1916 Race Street
Philadelphia, PA 19103

American Water Works Association (AWWA)
6666 West Quincy Avenue
Denver, CO 80235

American Welding Society (AWS)
P.O. Box 351040
Miami, FL 33135

Associated General Contractors of America Inc. (AGC)
1957 E Street, NW
Washington, DC 20006

Association of American Railroads (AAR)
American Railroad Building
50 F Street, NW
Washington, DC 20001

Chemical Manufacturers Association, Inc. (CMA)
2501 M Street, NW
Washington, DC 20037

Compressed Gas Association, Inc. (CGA)
1235 Jefferson Davis Highway
Arlington, VA 22202

Conveyor Equipment Manufacturers Association
932 Hungerford Drive, #36
Rockville, MD 20850

Factory Mutual Engineering Organization (FMEO)
1151 Boston-Providence Turnpike
Norwood, MA 02062

Federal Emergency Management Agency (FEMA)
Publications Office
500 C Street, SW
Washington, DC 20472

Health Physics Society (HPS)
1340 Old Chain Bridge Road
McLean, VA 22101

Human Factors Society (HFS)
P.O. Box 1369
Santa Monica, CA 90406

Hydraulic Institute
8 Sylvan Way
Parsippany, NJ 07054-3802

Industrial Health Foundation, Inc.
34 Penn Circle W
Pittsburgh, PA 15206

Industrial Risk Insurers (IRI)
85 Woodland Street
Hartford, CT 06102

Industrial Safety Equipment Association, Inc. (ISEA)
1901 North Moore Street
Arlington, VA 22209

Laser Institute of America (LIA)
12424 Research Parkway, Suite 130
Orlando, FL 32826

National Association of Manufacturers (NAM)
1331 Pennsylvania Avenue, NW, Suite 1500N
Washington, DC 20004

National Constructors Association (NCA)
1101 15th Street, NW
Washington, DC 20005

National Health Council, Inc.
350 Fifth Avenue, Room 1118
New York, NY 10018

National Fire Protection Association (NFPA)
Batterymarch Park
Quincy, MA 02269

National Propane Gas Association (NPGA)
1600 Eisenhower Lane
Lisle, IL 60532

National Safe Workplace Institute
122 South Michigan Avenue, Suite 1450
Chicago, IL 60603

National Safety Council (NSC)
1121 Spring Lake Drive
Itasca, IL 60143-3201

Power Tool Institute, Inc.
1095 Oceanview Drive
P.O. Box 818
Yachats, OR 97498-0818

Scaffolding, Shoring, and Forming Institute, Inc.
c/o Thomas Associates, Inc.
1230 Keith Building
Cleveland, OH 44115

Steel Plate Fabricators Association, Inc.
2400 South Downing Avenue
Westchester, IL 60153

Steel Tank Institute
570 Oakwood Road
Lake Zurich, IL 60047

Underwriters Laboratories Inc. (UL)
333 Pfingsten Road
Northbrook, IL 60062

7.3.3. Technical Societies

Many professional societies take an interest in workplace safety and are available as sources of information. A list of some of these organizations follows.

American Board of Industrial Hygiene
4600 West Saginaw, Suite 101
Lansing, MI 48917

American Chemical Society (ACS)
1155 16th Street, NW
Washington, DC 20036

American Conference of Governmental and Industrial Hygienists (ACGIH)
6500 Glenway Avenue, Bldg. D-7
Cincinnati, OH 45211

American Industrial Hygiene Association (AIHA)
P.O. Box 8390
345 White Pond Drive
Akron, OH 44320

American Institute of Chemical Engineers (AIChE)
345 East 47th Street
New York, NY 10017

American Institute of Mining, Metallurgical, and Petroleum Engineers (AIME)
345 East 47th Street
New York, NY 10017

American Public Health Association (APHA)
1015 15th Street, NW
Washington, DC 20005

American Society of Mechanical Engineers (ASME)
345 East 47th Street
New York, NY 10017

American Society of Safety Engineers (ASSE)
1800 East Oakton
Des Plaines, IL 60018

American Society for Training and Development (ASTD)
Box 1433, 1630 Duke Street
Alexandria, VA 22313

The Chlorine Institute
2001 L Street, NW
Washington, DC 20036

Institute of Electrical and Electronic Engineers (IEEE)
345 East 47th Street
New York, NY 10017

National Association of Suggestion Systems
230 North Michigan Avenue
Chicago, IL 60601

National Safety Management Society
3871 Piedmont Avenue
Oakland, CA 94611

7.4. REGULATORY PROGRAMS

7.4.1. Introduction

There are a number of reasons why accident prevention and safety programs should be an integral part of today's workplace. Following are six primary considerations:

- Needless destruction of life and health is morally unjustified.
- Failure to take necessary precautions against predictable accidents and occupational illnesses makes management and workers morally responsible for those accidents and illnesses.
- Accidents and occupational illnesses severely limit efficiency and productivity.
- Accidents and occupational illnesses produce far- reaching social harm.
- The safety movement has demonstrated that its techniques are effective in reducing accident rates and promoting efficiency.
- Recent legislation makes management responsible for safety and health in the workplace.

The last of these is the subject of the following sections. Section 7.4.2 gives an overview of the regulatory situation. Section 7.4.3 discusses some of the relevant legislation.

7.4.2. Overview

The high level of public interest in the control of hazardous substances will continue to generate new legislation.

In addition, prosecution of employers and safety and health professionals for negligence resulting in serious injury or accidental death of employees will become more common. As the safety profession becomes increasingly recognized, the public may hold some practitioners accountable for any omissions that may result in death or injury to workers in their organizations.

Because of the continued rapid change in regulations and the roles of the various agencies, it would not be productive to try to be comprehensive here. Management of any operation must become familiar and stay current with the situation. The unfortunate duplication and overlap of agency responsibilities, with its proliferation of reporting requirements, makes this more difficult.

Currently, there are perhaps three sets of regulations most pertinent to this volume:

- protection of personnel and society from hazardous substances (EPA)
- protection of plant personnel from all hazards of the workplace (OSHA)
- mandated safety analysis and continuing improvement of facilities (OSHA's PSM program)

The Environmental Protection Agency (EPA) had the original authority under the Clean Air Act (1967) and the Clean Water Act (1972). This authority has been broadened by amendments to those acts and by some of the newer legislation referred to in the next section. The EPA has authority through various acts and programs to:

- regulate actual discharges to the air and water
- reduce potential for discharges
- define acceptable emission control technology
- regulate methods of disposal for wastes
- collect and disseminate information on emissions
- define acceptable waste disposal technology

The OSHAct of 1970 created OSHA and authorized it to

- promulgate, modify, and revoke safety and health standards
- conduct inspections and investigations and issue citations, including proposed penalties
- require employers to keep records of safety and health data
- petition the courts to restrain imminent danger situations
- approve or reject state plans for programs under the Act
- provide training and education to employers and employees
- consult with employers, employees, and organizations on prevention of injuries and illnesses
- grant funds to the states for identification of program needs and for plan development, experiments, demonstrations, administration, and operation of programs

- develop and maintain a statistics program for occupational safety and health

The Process Safety Management (PSM) program of OSHA is described in 29CFR1910.119. It requires, among other things, that an employer with a process involving more than a defined quantity of a listed hazardous chemical perform initial hazard analyses on all such processes. The established schedule requires these to be complete in 1997. Acceptable methods of analysis are described in Section 7.1.1. The PSM rule also covers:

- *Employee participation.* This requires a written plan describing the development of the information required for PSM and makes all results available to employees.
- *Process safety information.* The rule requires a compilation of written safety information before beginning process hazard analysis. There is a detailed list of the minimum acceptable information, which must be available to all PHA participants. Use of this information can be limited by secrecy agreements.
- *Operating procedures.* These must be developed and implemented by the employer. The rule gives a detailed list of required subjects.
- *Training.* Initial training and refreshers at least every three years are required. Documentation must include some evidence that the employee has understood the training.
- *Contractors.* Employers and contractors have mutual responsibilities to share information and training.
- *Pre-startup review.* This is required for both new and modified facilities. It should cover design, procedures, employee training, and PHA.
- *Mechanical integrity.* This is aimed at a defined list of system components and covers quality assurance, maintenance programs and training, inspection and testing, and correction of deficiencies.
- *Hot work permits.* "Hot work" comprises welding, cutting, brazing, and similar flame- or spark-producing operations.
- Management of change. This covers the analysis, documentation, and training required before making any change in technology, procedures, or facility.
- *Incident investigation.* This is also the subject of Section 7.1.5. The emphasis is on prompt investigation and documented resolution.
- *Emergency planning and response.* There should be a program in accord with 29CFR1910.38(a).
- *Compliance audits.* These must be carried out at least every three years, and the employer must document a response and correct any deficiencies.

OSHA has more direct impact on day-to-day operation and on specifics of plant safety programs than does EPA. Administration of OSHA programs is through ten regional offices. Their addresses are given in Table 7-4.

TABLE 7-4 OSHA Regional Offices
Region I (Connecticut, Maine, Massachusetts, New Hampshire, Rhode Island, Vermont) 133 Portland Street, 1st Floor Boston, MA 02114 (617) 565-7164
Region II (New York, New Jersey, Puerto Rico, Virgin Islands, Canal Zone) 201 Varick Street—Room 670 New York, NY 10014 (212) 337-2378
Region III (Delaware, District of Columbia, Maryland, Pennsylvania, Virginia, West Virginia) Gateway Building, Suite 2100 3535 Market Street Philadelphia, PA 19104 (215) 596-1201
Region IV (Alabama, Florida, Georgia, Kentucky, Mississippi, North Carolina, South Carolina, Tennessee) 1375 Peachtree Street NE, Suite 587 Atlanta, GA 30367 (404) 347-3573
Region V (Illinois, Indiana, Michigan, Minnesota, Ohio, Wisconsin) 230 South Dearborn Street, Room 3244 Chicago, IL 60604 (312) 353-2220
Region VI (Arkansas, Louisiana, New Mexico, Oklahoma, Texas) 525 Griffin Street, Room 602 Dallas, TX 75202 (214) 767-4731
Region VII (Iowa, Kansas, Missouri, Nebraska) 911 Walnut Street, Room 406 Kansas City, MO 64106 (816) 426-5861
Region VIII (Colorado, Montana, North Dakota, South Dakota, Utah, Wyoming) Federal Building, Room 1576 1961 Stout Street Denver, CO 80204 (303) 844-3061
Region IX (Arizona, California, Hawaii, Nevada, Guam, American Samoa, Trust Territory of the Pacific Islands) 71 Stevenson Street, Suite 415 San Francisco, CA 94105 (415) 744-6670
Region X (Alaska, Idaho, Oregon, Washington) 1111 Third Avenue, Suite 715 Seattle, WA 98101 (206) 553-5930

7.4.3. Examples

This section lists some of the legislation and regulatory bodies that have had a major impact on process plant design and operation. It is impossible to treat state or local requirements in any way. The plant operator must of course see to it that he abides by all these, in addition to any specific insurance requirements. The following list will provide some guidance and explain some of the many acronyms that apply to federal agencies and national programs.

CAA(A)—**Clean Air Act (Amendments).**
The original act established EPA's jurisdiction over air quality. Amendments have resulted in NAAQS, NESHAPS, the list of regulated air toxic chemicals, and a host of other air quality regulations.

CAER—**Community Awareness and Emergency Response.**
This requires sharing of safety-related information and emergency planning with the public ("right-to-know" legislation). Part of SARA Title III.

CERCLA—**Comprehensive Environmental Response, Compensation, and Liability Act of 1980.**
This is "Superfund." It deals with identification and remediation of abandoned hazardous waste sites.

HCS—**Hazard Communication Standard.**
A standard established by OSHA which requires a regulated facility to establish a system for communication of risks and hazardous properties of materials to employees.

HMR—**Hazardous Materials Regulations.**
DOT regulations dealing with the transport of hazardous materials.

HSWA—**Hazardous and Solid Waste Amendments.**
Amendments to RCRA in 1984 which tightened requirements on operating plants and set the timetable for land bans.

LEPC—**Local Emergency Planning Committee.**
Not a regulatory agency as such. LEPC's are groups mandated by SARA that are responsible for planning responses to emergencies. They include personnel from industry, emergency responders, and the community. Requirements vary from state to state.

RCRA—**Resource Conservation and Recovery Act.**
This regulates wastes currently being generated and promotes recovery or reduction. It also covers treatment, storage, disposal, and transportation of wastes.

SARA—Superfund Amendment and Reauthorization Act.
This reauthorized CERCLA and expanded government jurisdiction.

SARA Title III—Section of SARA.
It requires industry to develop comprehensive emergency response plans and mandates public disclosure of hazards of materials handled or stored in certain quantities. The "Superfund sites" are those officially listed locations that contain hazardous wastes and that are to be remediated under CERCLA. They are identified on the National Priority List (NPL).

Superfund—See CERCLA and SARA.

TSCA—Toxic Substances Control Act.
This requires disclosure of information on toxicity of new materials before entering commercial production.

REFERENCES

1. Kletz, T. A., *Hazop and Hazan—Identifying and Assessing Process Industry Hazards*, 3d ed., Institution of Chemical Engineers, Rugby, UK, 1992.
2. Crowl, D. A., and Louvar, J. F., *Chemical Process Safety: Fundamentals with Applications*, Prentice-Hall, Englewood Cliffs, NJ, 1990.
3. O'Brien, T. F., "Emergency Vent Scrubbing Systems—Design; Operation; Hazard Analysis," Electrode Corp. Chlorine/Chlorate Seminar, Cleveland, 1991.
4. National Fire Codes, including
   a. Building Officials and Code Administrators International (BOCA): *Basic Fire Prevention Code*.
   b. International Conference of Building Officials (ICBO): *Uniform Fire Codes*.
   c. Southern Building Code Congress International (SBCCI): *Standard Fire Prevention Code*.
   d. National Fire Protection Association (NFPA): NFPA 1, *Fire Prevention Code*.
5. Fawcett, H. H., and Wood, W. J., *Safety and Accident Prevention in Chemical Operations*, Interscience, New York, 1965.
6. National Safety Council, *Accident Prevention Manual for Business and Industry*, volume subtitled *Administration and Programs* and volume subtitled *Engineering and Technology*, 10th ed., NSC, Itasca, IL, 1992.
7. Olishifski, J. B., ed. *Fundamentals of Industrial Hygiene*, 2d ed., National Safety Council, Itasca, IL, 1979.
8. *Chem. Week*, Special Issue, 17 July 1991.
9. *Chem. Week*, Special Issue, 8 December 1993.

BIBLIOGRAPHY

Air Movement and Control Association, Publication 99-86, *Standards Handbook,* AMCA, Arlington Heights, IL, 1986.

Air Movement and Control Association, Publication 202-88, Troubleshooting, AMCA, Arlington Heights, IL, 1988.

Air Movement and Control Association, Publication 410-90, *Recommended Safety Practices for Users and Installers of Industrial and Commercial Fans* AMCA, Arlington Heights, IL, 1990.

Alerich, W. N., *Electric Motor Controls*, 4th ed., Delmar Publishers, Albany, NY, 1988.

American Conference of Governmental and Industrial Hygienists, *Guide to Occupational Exposure Values*, ACGIH, Cincinnati, 1992.

American Conference of Governmental and Industrial Hygienists, *1992–1993 TLVs for Chemical Substances and Physical Agents*, ACGIH, Cincinnati, 1992.

American Institute of Chemical Engineers, *AIChE Pump Manual*, Section 3—Definitions of Pump Classifications.

American Petroleum Institute, Standard 612—Special Purpose Steam Turbines for Refinery Service, 3d ed., API, Washington, 1987.

American Petroleum Institute, Standard 611—General Purpose Steam Turbines for Refinery Service, 3d ed., API, Washington, 1988.

American Society of Mechanical Engineers, PTC 6-1976, Steam Turbines, Performance Test Codes, ASME, New York, 1976.

American Society for Testing and Materials, *Fire Safety Science and Engineering*, ASTM, Philadelphia, 1985.

Avallone, E. A., and T. Baumeister, eds. *Marks' Standard Handbook for Mechanical Engineers*, 9th ed., McGraw-Hill, New York, 1987.

Benedetti, R. P., ed. *Flammable and Combustible Liquids Code Handbook*, 4th ed., NFPA, Quincy, MA, 1987.

Bierlein, L. W., *Red Book on Transportation of Hazardous Materials*, Cahners Books Intl., Boston, 1977.

Bodurtha, F. T., *Industrial Explosion Prevention and Protection*, McGraw-Hill, New York, 1980.

Bolton, W., *Engineering Materials Pocket Book*, CRC Press, Boca Raton, FL, 1989.

Bouchard, J. K., ed. *Automatic Sprinkler Systems Handbook*, 3d ed., National Fire Protection Association, Quincy, MA, 1987.

Bretherick, L., *Bretherick's Handbook of Reactive Chemical Hazards*, 4th ed., Butterworths, London, 1990.

Brown, R. N., *Compressors: Selection & Sizing*, Gulf Publishing Company, Houston, 1986.

Brunner, C. R., *Handbook of Hazardous Waste Incineration*, TAB Books, Blue Ridge Summit, PA, 1989.

Buschart, R. J., *Electrical and Instrumentation Safety for Chemical Processes*, Van Nostrand Reinhold, New York, 1991.

Cheremisinoff, P. N., and R. A. Young, *Pollution Engineering Practice Handbook*, Ann Arbor Science Publishers, Ann Arbor, MI, 1976.

Colijn, H., *Mechanical Conveyors for Bulk Solids*, Elsevier, New York, 1985.

Cook, E. M., and H. D. DuMont, *Process Drying Practice*, McGraw-Hill, New York, 1991.

Corripio, A. B., *Tuning of Industrial Controls*, Instrument Society of America, Research Triangle Park, NC, 199.

Cote, A. E., ed. *Fire Protection Handbook*, 16th ed., NFPA, Quincy, MA, 1986.

Craig, R. L., ed. *Training and Development Handbook*, McGraw-Hill, New York, 1976.

Cralley, L. V., and L. J. Cralley, *Industrial Hygiene Aspects of Plant Operations*, Macmillan, New York, 1985.

Davis, M. L., and D. A. Cornell, *Introduction to Environmental Engineering*, 2d ed., McGraw-Hill, New York, 1991.

Department of Labor, *Training Requirements in OSHA Standards and Training Guidelines* (OSHA 2254).

Farrall, A. W., ed. *Food Engineering Systems, Volume 1—Operations*, Chapter 24, Sanitation and Water Supply, AVI Publishing Company, Westport, CT, 1976.

Fawcett, H. A. and W. S. Wood, eds. *Safety and Accident Prevention in Chemical Operations*, 2d ed., John Wiley, New York, 1982.

Fink, D. G., and H. W. Beatty, eds. *Standard Handbook for Electrical Engineers*, 13th ed., McGraw-Hill, New York, 1993.

Fisher, T. G., *Alarm and Interlock Systems*, Instrument Society of America, Research Triangle Park, NC, 1984.

Fisher, T. G., *Batch Control Systems: Design, Application, and Implementation*, Instrument Society of America, Research Triangle Park, NC, 1984.

Giachino, J. W., W. Weeks, and G. S. Johnson, *Welding Technology*, 2d ed., American Technical Publishers, Homewood, IL, 1973.

Haselden, G. G., ed. *Cryogenic Fundamentals*, Academic Press, New York, 1971.

Hunter, R. P., *Automated Process Control Systems: Concepts and Hardware*, Prentice-Hall, Inc., Englewood Cliffs, NJ, 1987.

Hydraulic Institute, *Engineering Data Book*, Hydraulic Institute, Cleveland, 1990.

Institution of Chemical Engineers, *Hazards XI, New Directions in Process Safety*, Symposium Series No. 124, Hemisphere, New York, 1991.

Instrument Society of America, *Standards and Recommended Practices for Instrumentation and Control*, 11th ed., 3 vols., ISA, Research Triangle Park, NC, 1991.

Karassik, I. G., et al, eds., *Pump Handbook*, 2d ed., McGraw-Hill, New York, 1986.

M. W. Kellogg, Inc., *Design of Piping Systems*, John Wiley, New York, 1968.

Kissell, T. E, *Modern Industrial/Electric Motor Controls, Operation, Installation, and Troubleshooting*, Prentice-Hall, Englewood Cliffs, NJ, 1990.

Kletz, T. A,, *Lessons from Disaster: How Organizations Have No Memory and Accidents Recur*, Gulf Publishing Co., Houston, 1993.

Kuelling, H., and K. Hammer, *Explosion Hazards When Processing Pharmaceutical Products in Fluid-Bed Equipment*, internal publication, Aeromatic AG, Bubendorf (Switzerland), 1987.

Kusko, A., *Emergency/Standby Power Systems*, McGraw-Hill, New York, 1987.

Lee, R. H., "Lightning Protection of Buildings", IEEE Transactions of Industrial Applications, May 1979.

Lewis, R. J., Sr., *Sax's Dangerous Properties of Industrial Materials*, 8th ed., 3 vols., Van Nostrand Reinhold, New York, 1992.

Lide, D. R., ed., *CRC Handbook of Chemistry and Physics*, 72d ed., CRC Press, Boca Raton, FL, 1992.

Lindberg, R. A., *Processes and Materials of Manufacture*, 2d ed., Allyn and Bacon, Boston, 1977.

Magison, E. C., *Electrical Instruments in Hazardous Locations*,, 3d ed., Instrument Society of America, Research Triangle Park, NC, 1978.

Magison, E. C., *Intrinsic Safety*, Instrument Society of America, Research Triangle Park, NC, 1984.

McGraw-Hill *Encyclopedia of Science and Technology*, 5th ed., McGraw-Hill, New York, 1982.

Mecklenburgh, J. C., *Process Plant Layout*, John Wiley & Sons, New York, 1985.

Meyer, E., *Chemistry of Hazardous Materials*, Prentice-Hall, Englewood Cliffs, NJ, 1977.

Michaels, E. C., "Grounding and bonding in hazardous locations," *Plant Engineering*, 35, 17 Sept., 1 Oct., and 23 Dec. 1981.

Miller, R. K., *Handbook of Industrial Noise Management*, Fairmont Press, Atlanta, 1976.

Miller, R. K., and W. V. Montone, *Handbook of Acoustical Enclosures and Barriers*, Fairmont Press, Atlanta, 1978.

Moore, R. E., "Selecting Materials To Meet Environmental Conditions," *Chem. Eng. 86*, No. 14, 101–103, 2 July 1979.

Moore, R. E., "Selecting Materials To Resist Corrosive Conditions," *Chem. Eng. 86*, No. 16, 91–94, 30 July 1979.

Moran, R. D., *OSHA Handbook*, Government Institutes, Inc., Rockville, MD, 1987.

Moyer, C. A., and M. A. Francis, *Clean Air Act Handbook*, Clark Boardman Co., New York, 1991.

National Electrical Manufacturers Association, Publication No. SM21—Multistage Steam Turbines for Mechanical Drive Service, NEMA, Washington, 1970.

National Electrical Manufacturers Association, Publication No. SM23—Steam Turbines for Mechanical Drive Service, NEMA, Washington, 1986.

National Fire Protection Association, *Flammable and Combustible Liquids Code Handbook*, 4th ed., NFPA, Quincy, MA, 1991.

National Fire Protection Association, *Life Safety Code Handbook*, 5th ed., NFPA, Quincy, MA, 1991.

National Fire Protection Association (NFPA) Codes, Standards, Recommended Practices, and Manuals:

> NFPA 13: Standard for the Installation of Sprinkler Systems.
> NFPA 13A: Recommended Practice for the Inspection, Testing and Maintenance of Sprinkler Systems.

NFPA 15: Standard for Water Spray Fixed Systems for Fire Protection.

NFPA 43A: Code for the Storage of Liquid and Solid Oxidizing Materials.

NFPA 49: Hazardous Chemicals Data.

NFPA 61A: Standard for Prevention of Fire and Dust Explosions in Facilities Manufacturing and Handling Starch.

NFPA 61B: Standard for the Prevention of Fires and Explosions in Grain Elevators and Facilities Handling Bulk Raw Agricultural Commodities.

NFPA 68: Guide for Explosion Venting.

NFPA 69: Standard for Explosion Prevention Systems.

NFPA 72E: Standard on Automatic Fire Detectors.

NFPA 77: Recommended Practice on Static Electricity.

NFPA 85A: Standard for Prevention of Furnace Explosions in Fuel Oil- and Natural Gas-Fired Single Burner Boiler-Furnaces.

NFPA 85B: Standard for Prevention of Furnace Explosions in Natural Gas-Fired Multiple Burner Boiler-Furnaces.

NFPA 85D: Standard for Prevention of Furnace Explosions in Fuel Oil-Fired Multiple Burner Boiler-Furnaces.

NFPA 85E: Standard for Prevention of Furnace Explosions in Pulverized Coal-Fired Multiple Burner Boiler-Furnaces.

NFPA 85G: Standard for Prevention of Furnace Implosions in Multiple Burner Boiler-Furnaces.

NFPA 85H: Standard for Prevention of Combustion Hazards in Atmospheric Fluidized Bed Combustion System Boilers.

NFPA 86: Standard for Ovens and Furnaces.

NFPA 86C: Standard for Industrial Furnaces Using a Special Processing Atmosphere.

NFPA 86C: Standard for Industrial Furnaces Using Vacuum as an Atmosphere.

NFPA 91: Installation of Blower, Exhaust Systems for Dust, Stock and Vapor Removal or Conveying.

NFPA 231: Standard for Indoor General Storage.

NFPA 231C: Standard for Rack Storage of Materials.

NFPA 321: Basic Classification of Flammable and Combustible Liquids.

NFPA 325M: Fire Hazard Properties of Flammable Liquids, Gases, and Volatile Solids.

NFPA 327: Standard Procedures for Cleaning or Safeguarding Small Tanks and Containers.

NFPA 491: Manual of Hazardous Chemical Reactions.

NFPA 505: Fire Safety Standard for Powered Industrial Trucks Including Type Designations, Areas of Use, Maintenance and Operations.

NFPA 650: Standard for Pneumatic Conveying Systems for Handling Combustible Materials.

NFPA 654: Standard for the Prevention of Fire and Dust Explosions in the Chemical, Dye, Pharmaceutical, and Plastics Industry.

NFPA 655: Standard for the Prevention of Sulfur Fires and Explosions.

NFPA 704: Standard System for the Identification of the Fire Hazards of Materials.

National Safety Council, *Accident Prevention Manual for Industrial Operations*, 6th ed., NSC, Chicago, 1969.

Nayyar, M. L., ed. *Piping Handbook*, 6th ed., McGraw-Hill, New York, 1992.

Neglia, J. P., "Personal Protection for Hazardous Conditions," The National Environmental Journal, March/April 1992.

Noyes, R., ed. *Handbook of Leak, Spill and Accidental Release Prevention Techniques*, Princeton Technical Publishers, Lexington, MA, 1992.

Nuclear Regulatory Commission, *NRC Rules and Regulations* (10CFR, Chap. 1), Washington, 1992.

Occupational Safety and Health Administration, *OSHA Compliance Encyclopedia*, 3 vols., Business and Legal Reports, Inc., Madison, CT, 1993.

Perry, R. H., and D. W. Green, eds. *Perry's Chemical Engineers' Handbook*, 6th ed., McGraw-Hill, New York, 1984

Phifer, R. W., and W. R. McTigue, Jr., *Waste Management for Small Quantity Generators*, Lewis Publishers, Chelsea, MI, 1988.

Plog, B. A., ed. *Fundamentals of Industrial Hygiene*, 3d ed., National Safety Council, Chicago, 1988.

Sanders, M. S., and E. J. McCormick, *Human Factors in Engineering and Design*, 7th ed., McGraw-Hill, New York, 1993.

Sanders, R. A., *Management of Change in Chemical Plants*, Butterworth-Heinemann, Oxford, 1993.

Sandler, H. J. and E. T. Luckiewicz, *Practical Process Engineering, a working approach to plant design*, McGraw-Hill, New York, 1987.

Scheflan, L., and M. B. Jacobs, *Handbook of Solvents*, Van Nostrand, New York, 1953.

Schweitzer, P. A., *Handbook of Corrosion Resistant Piping*, Industrial Press, New York, 1965

Schweitzer, P. A., *Corrosion Resistance Tables*, 2d ed., Marcel Dekker, New York, 1986.

Sherwood, D. R., and D. J. Whistance, *The Piping Guide*, Syentek Books, Cotati, CA, 1973.

Smith, P. R., and T. J. Van Laan, *Piping and Pipe Support Systems, Design and Engineering*, McGraw-Hill, New York, 1987.

Spotts, M. F., *Design of Machine Elements*, 5th ed., pp. 601–607, Prentice-Hall, Englewood Cliffs, NJ, 1978.

Thielsch, H., *Defects & Failures in Pressure Vessels and Piping*, Reinhold, New York, 1962.

Weiss, G., *Hazardous Chemicals Data Book*, 2d ed., Princeton Technical Publishers, Lexington, MA, 1986.

White, I. F., "Single-Medium Systems: The Answer to Heating/Cooling Cycle Problems", Process Eng., November 1988.

Williams, P. L. and J. L. Burson, eds. *Industrial Toxicology, Safety and Health Applications in the Workplace*, Van Nostrand Reinhold, New York, 1985.

Young, J. A., *Improving Safety in the Chemical Laboratory: A Practical Guide*, Wiley-Interscience, New York, 1991.

Zappe, R. W., *Valve Selection Handbook*, 2d ed., Gulf Publishing Co., Houston, 1987.

INDEX